EARTH BEFORE THE DI

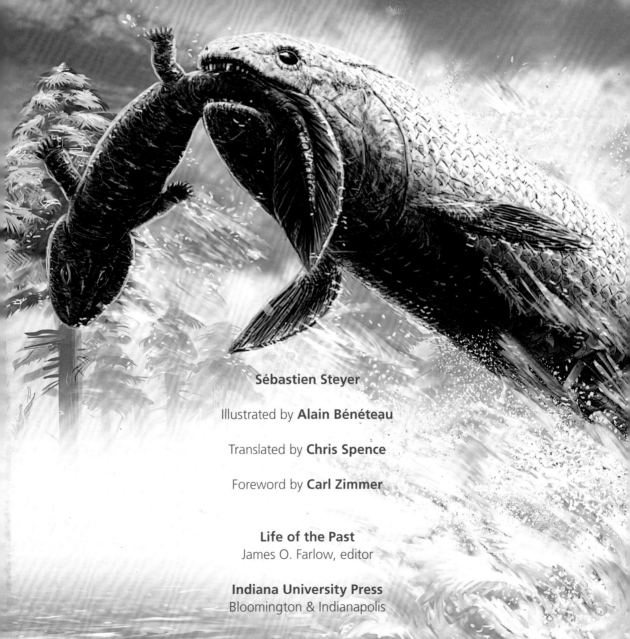

EARTH BEFORE THE DINOSAURS

Sébastien Steyer

Illustrated by **Alain Bénéteau**

Translated by **Chris Spence**

Foreword by **Carl Zimmer**

Life of the Past
James O. Farlow, editor

Indiana University Press
Bloomington & Indianapolis

This book is a publication of

Indiana University Press
601 North Morton Street
Bloomington, Indiana 47404-3797 USA

iupress.indiana.edu

Telephone orders 800-842-6796
Fax orders 812-855-7931

Originally published as La Terre avant les dinosaures © Éditions Belin-Paris, 2009

English Translation © 2012 by Indiana University Press

Published with the support of the French Cultural Ministry, National Center for the Book Ouvrage publié avec le concours du Ministère français chargé de la culture – Centre national du livre

All rights reserved

No part of this book may be reproduced or utilized in any form or by any means, electronic or mechanical, including photocopying and recording, or by any information storage and retrieval system, without permission in writing from the publisher. The Association of American University Presses' Resolution on Permissions constitutes the only exception to this prohibition.

⊖ The paper used in this publication meets the minimum requirements of the American National Standard for Information Sciences – Permanence of Paper for Printed Library Materials, ANSI Z39.48-1992.

Manufactured in China

Library of Congress Cataloging-in-Publication Data

Steyer, Sebastien.
 [Terre avant les dinosaures. English]
 Earth before the dinosaurs / Sébastien Steyer ; illustrations and figures by Alain Bénéteau ; translated from the French by Chris Spence.
  p. cm. – (Life of the past)
 Includes bibliographical references and index.
  ISBN 978-0-253-22380-7 (pbk. : alk. paper) 1. Vertebrates – Evolution.
 2. Extremities (Anatomy) – Evolution.
 3. Vertebrates, Fossil.
 4. Paleontology – Paleozoic. I. Title.
  QL607.5.S7413 2012
  596–dc23

2011036525

1 2 3 4 5 17 16 15 14 13 12

To Thelma and Muriel

Alice, Agata, Mathieu, and Christine

CONTENTS

FOREWORD BY CARL ZIMMER > IX >

PREFACE – AND A SHORT PREAMBLE > XI >

ACKNOWLEDGMENTS > XIII >

1 The Great Transition > 1 >

2 Limbs: How Do They Work? > 43 >

3 The Pangaean Chronicles > 59 >

4 Between Earth and Sky > 111 >

5 A Brief Guide to Paleontology > 161 >

BIBLIOGRAPHY > 175 >

INDEX > 179 >

FOREWORD

THE NEW YORKER HAS COMMISSIONED MANY FAMOUS COVERS, but the most famous of all was the one that appeared on the March 29, 1976 issue. *View of the World from 9th Avenue* is a charming work of ink, pencil, and watercolor by the artist Saul Steinberg. It is best viewed from bottom to top. At the bottom is Ninth Avenue, where you can easily make out the cars, streetlights, pedestrians, and a red and yellow sign in front of a lot that reads PARK. Your eye moves up a cross street to Tenth Avenue, and from there to the Hudson River, traversing the middle of the cover with an Amazonian majesty. Beyond the river is a grimy streak simply labeled Jersey. Beyond Jersey lies a non-descript block of green with three scattered mountains shaped like lumps of clay, along with a handful of names, like Kansas City, Chicago, and Utah. On the other side of the United States, the Pacific Ocean looks about the same size as the Hudson. The eye reaches the top of the picture at last, where there are three low mounds, marked China, Japan, and Russia.

Steinberg pokes fun at the mix of self-importance and parochialism common among Manhattanites. But, like all good cartoons, *View of the World from 9th Avenue* speaks to something deeper, a trait common beyond New York's borders. All of us perceive our immediate surroundings as the center of the world; further away, our mind's eye gets blurry; and most of the world shrinks away to a minor abstraction.

This rule holds not just for space, but also for time. For Americans, the history of the United States is colossal, a rich expanse of triumphs and disaster, of revolutions both military and social. Before 1776, our historical eye gets fuzzy. We look back at the Middle Ages as a blur of cathedrals and alchemy. The empires of Assyria and Babylon and Egypt are barely visible. And before civilization, there isn't much at all. If Steinberg were to paint *View of History from the Twenty-First Century*, the farthest reaches of the picture would only be marked by a few dinosaurs.

It's understandable that dinosaurs should dominate our view of the distant past. When we visit natural history museums, it's the dinosaurs that tower over our heads. Or, at least, it's the big dinosaurs that tower over our heads. Of course, many mid-size and petite dinosaurs existed during as well, which are no less interesting for their smaller bulk. In fact, some of the most interesting of all were the very smallest: the minuscule theropods that evolved feathers and gave rise to birds, the only dinosaurs still alive today.

Beyond the dinosaur horizon are many more unfairly neglected vertebrates. The early tetrapods that came on land, for example, or their descendants that began to glide through the air or which returned to the water. Their bodies plunged off into all manner of surreal extremes: serpentine trunks, heads like the cow catchers on trains, bodies with ribs stretched out into fans. *Earth before the Dinosaurs* is a wonderful way to add this distant period to our mental view of history. It's not just a history of strange, extinct lineages of amphibians and reptiles, however. Some of the strangest of the animals, which look vaguely like saber-toothed turtles, are our own ancient relatives, on the lineage that would eventually lead to mammals. The view may be distant, but it's also personal.

Carl Zimmer
Guilford, CT
January 9, 2012

PREFACE – AND A SHORT PREAMBLE

DINOSAURS ARE THE UNCONTESTED STARS OF PALEONTOLOGY. Given their often spectacular morphology and extinction (or, rather, partial extinction—as birds are dinosaurs), when the Earth was struck by an enormous meteorite and immense volcanic eruptions shook the globe, they hog the limelight to such a degree that we often forget that before them there were other animals. To repair this injustice, this book invites you on a journey to a time before the dinosaurs—into a distant past to discover animals as surprising as they are fascinating, back to a time when our planet experienced the worst of all the mass extinctions it has ever seen (the famous life crises when life nearly disappeared from Earth). On the way we will discover how some commonly held beliefs on the evolution of species are completely false.

Of course we will not attempt to map out Earth's biodiversity from its very beginnings (approximately between 3 and 3.5 *billion* years ago) to the age of the dinosaurs (from about 200 million years ago), but rather focus our objective on animals which, like the dinosaurs (and also like amphibians, "reptiles," and mammals) possess backbones and four limbs equipped with digits: the vertebrate tetrapods. We will begin by probing into the origins of these tetrapods. For a very long time the first tetrapods have been associated with an emergence from the water: just as in human birth, after slowly developing enveloped in a protective liquid (the amniotic fluid), the baby leaves the womb and its liquid environment—so must four-limbed animals have left the water, once their brand new limbs had been cobbled together by evolution. In other words, we believed the early tetrapods were terrestrial and that their limbs served them to get about on land. Thus, with the advent of the tetrapods, vertebrates had, it was thought, "succeeded in leaving the water."

This version of events may make for a good story, but is it true?

To answer this question we will visit the estuaries and deltas of our planet around 370 million years ago, where numerous surprises await (chapter 1). By close observation of limb morphogenesis during embryonic development we will better understand the evolutionary processes at work at the origin of the tetrapods (chapter 2). After this we will delve into the tetrapods' past and get a taste of their incredible diversity (chapters 3 and 4). Whether in flowing water, seas, or in the heart of thick forest, we will encounter forms with varied morphologies and lifestyles—sometimes of an appearance so strange that they may seem like creatures from a science fiction novel.

In our wanderings we will come across numerous fossils. How do paleontologists bring them back to life? That is to say, how do these Sherlock Holmeses of evolution piece together fossil clues—often incomplete—to recreate the morphology and lifestyle of an animal? This question will be the subject of the last chapter, which will provide a glimpse into the toolbox at the disposal of the twenty-first-century paleontologist.

This book does not pretend to be an exhaustive work on what is a vast and multidisciplinary subject—the story of the early tetrapods. But I hope that by leafing through these pages you will also discover a new world, a little-known continent far removed from the familiar haunts of the iconic dinosaurs or the well-trampled world of pre-humans—much as Darwin did when he first set foot on the Galapagos Archipelago. This new world is that of the Earth before the dinosaurs, a place that deserves a visit and one to which I have had the privilege of devoting my career. May this modest book intrigue and—who knows?—perhaps transmit the paleontology bug to its readers. As I hope you will discover, it is highly infectious!

A Short Preamble

To find their bearings in the incredible diversity that characterizes life on Earth, scientists attempt to reconstitute the lineages between organisms and discover who is most closely related to whom. They classify organisms, living and fossil, into different groups that reflect their evolution as accurately as possible. This classification is known as phylogeny. The groups it comprises are known as clades or taxa; these in turn comprise all living beings that share a common ancestor that is unique and exclusive to each group. We identify the members of a clade by characteristics passed on by this common ancestor that are unique to them. Thus, the common ancestor unique to all vertebrates bequeathed vertebrae to all its descendants. An animal that possesses vertebrae is assigned to the clade of vertebrates. Within the vertebrate clade, the common ancestor of tetrapods passed on limbs with digits to all its descendants (although some, like snakes, have since lost theirs). A vertebrate that possesses such limbs will be assigned to the clade of tetrapods.

Several traditional groups have disappeared from phylogenetic classification: these are known as grades. Fish form a grade but a fish like the lungfish or the coelacanth is more closely related to a tetrapod (human beings, for example, are tetrapods) than to other fish such as trout. Reptiles also traditionally form a grade that, in the living world, comprises turtles, lizards, snakes, and crocodiles. There is no specific characteristic that would have been passed on by a common ancestor to all these animals; the common ancestor of all reptiles is also that of birds and mammals (in other words, all the amniotes). Nevertheless, reptiles in the modern sense (the sauropsids), form a clade, but it includes the birds. For ease of reading, grades are not placed in quotation marks in this book.

ACKNOWLEDGMENTS

NUMEROUS COLLEAGUES HAVE HELPED ME WITH ADVICE AND suggestions, and I would like to thank the following people for their support—without which this book would not have seen the light of day: Vincent Dupret, Jean-Claude Rage, Dino Frey, Roger Malcom Smith, Ken Angielczyk, Christian Sidor, Stuart Sumida, Nour-Eddine Jalil, Silvio Renesto, Georges Gand, Ted Daeschler, Jennifer Clack, Olivier Béthoux, Brigitte Meyer-Berthaud, Marcello Pettineo, Herbert Thomas, Gäel Clément, Damien Germain, Jocelyn Falconnet, and Sophie Léonard—not to mention the many other colleagues cited in the descriptions and photographs. A big thank you also to Sophie Albouy for the indispensable task of proofreading.

This book would not be what it is without Alain Bénéteau, whose illustrations have brought back to life dearly beloved but sadly vanished organisms. As paleontological excavations play such an important part in this work, I must also thank those who are always ready to roll up their sleeves and get dirty: Georges Gand, Sophie Hervet, Suzanne Jiquel, Bernard Blanc, Chris Spence, Gaël De Ploeg, Jacques Rival, Pierre Bleyzac, Renaud Vacant, and Jacques Garric. If I have forgotten anyone, please accept my humble apologies.

Thanks to Indiana University Press for believing in this project.

This book was written in Paris, Saint-Paul-Trois-Châteaux (Drôme, France), Gigouzac (Lot, France), and Arlit and Agadez (Niger). These pages are full of happy memories of these places.

EARTH BEFORE THE DINOSAURS

THE GREAT TRANSITION 1

THE STORY OF THE EARLIEST FOUR-LEGGED VERTEBRATES, the early tetrapods, has been rewritten in the last few decades. As we saw in the preface, their beginnings were once inextricably linked to the emergence of life from the water. To understand the reasons for this misconception, let us look at land vertebrates today. They are all tetrapods—that is to say, vertebrates that possess (or their ancestors possessed) chiridian limbs (limbs with digits; fig. 1.1). The clawed paw of the bear or the muscular legs of the horse immediately spring to mind, but the wing of a parrot is also a chiridian limb comprising three parts, including the characteristic finger bones. Even if it is not immediately obvious, snakes are also tetrapods. Although their legs are conspicuously absent, their ancestors had them.

Although all terrestrial vertebrates are tetrapods, out of the 30,000 species of amphibians, sauropsids (birds and reptiles), and mammals that make up the tetrapods of today, not all are land based—far from it. Take dolphins and whales: their "fins" are chiridian limbs. As for amphibians, their connection to the water is evident. Despite this, the appearance of the first chiridian limbs (and therefore the early tetrapods) around 370 million years ago (also written as 370 Ma) was, until the 1990s, associated with the conquest of the land. It was believed that the first chiridian limbs served the tetrapods as a means of getting about on terra firma. This belief is completely false and a prime example of our anthropocentric view of the world.

That a human (a terrestrial animal, needless to say) walks with its chiridian limbs is not reason to declare that their limbs' initial evolutionary function was to facilitate walking. For one thing, structures do not necessarily emerge to fulfill a single function; also, in the natural world today, a function associated with a structure often differs from the function (or functions) that same structure served in the past.

We must remember that the notion of "conquest" has no place in evolution because it assumes that nature has some predetermined goal—or, put another way, that evolution follows trends and directions. This is certainly not the case. It is humans who define evolutionary trends, more often than not while contemplating themselves in the mirror. It is easy to understand how this "modest" primate of dreamy inclination has interpreted the "path of evolution" as culminating inevitably at its own door. Taken a step further, colonization of the skies would represent some sort of ultimate evolutionary goal for our own species, and the early tetrapods intrepid pioneers on a mission to colonize the land. This is far from the

> After long reflection, I cannot avoid the conviction that no innate tendency to progressive development exists.
>
> **Charles Darwin (1809–1882)**

Rhizodus, an ichthyan sarcopterygian of up to 8 meters in length from the Devonian, attacks *Acanthostega*. Although the latter possessed chiridian limbs (it was one of the earliest tetrapods), it was aquatic.

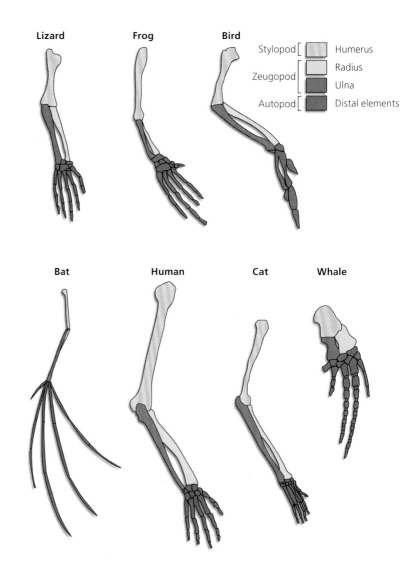

1.1. The forelimbs of tetrapods. These are all chiridian limbs—in other words, the most distal segment (the autopod) is equipped with digits. This structure is characteristic of tetrapods: whether walking, running, swimming or jumping, they share this character inherited from a unique, common ancestor. The chiridian limb is composed of three elements: the stylopod (humerus for the forelimb, femur for the hindlimb), the zeugopod (radius and ulna for the forelimb, tibia and fibula for the hindlimb), and the autopod (comprising the wrist or ankle bones, hands or feet, and the digits).

truth. Evolution has no endgame. The chiridian limb was not "invented" to conquer the land, and the exodus of vertebrates from the water was neither an exploit nor a success. It was pure chance—one of a multitude of avatars in the evolution of life on Earth. In fact, the earliest tetrapods in the fossil record, *Ichthyostega* and *Acanthostega* (370 Ma), were aquatic!

The teleological vision of tetrapod land conquest, which explains a phenomenon by its end result, persists to this day, merely mutating with time. The period that saw the early tetrapods emerge from the water—some 330 million years ago—well after the first appearance of the chiridian limb, is smoothly presented as "shedding the chains" of an aquatic environment. Is terrestrial locomotion better than swimming? Is a terrestrial tetrapod superior to an aquatic vertebrate, tetrapod or otherwise? The answer to both questions is a negative. A terrestrial organism is neither superior nor more evolved than an aquatic one; it is different, and that is all. A horse is not more "evolved" than a salmon, and possessing limbs—whether limbs, paddles, or wings—is no more remarkable than having fins. "Old does not mean hidebound in a Darwinian world," as paleontologist Stephen Jay Gould once put it (1991: 277). The wings of a hummingbird (a tetrapod), like the fins of a manta ray (a non-tetrapod), are both capable of the most masterful acrobatic displays.

To try to understand the earliest tetrapods, in this first chapter we will take a gentle stroll around the tree of life. We will take a look at creatures, both living and dead, closely related to the tetrapods, and trace the emergence of the innovation that allowed vertebrates to colonize the land, the first chiridian limb – an invention patented 40 million years earlier in the water!

The Coelacanth: Did You Say "Living Fossil"?

It is difficult not to start any meaningful history of the tetrapods without first discussing one of its closest living relatives, the (perhaps all too) legendary coelacanth (fig. 1.2). A coelacanth, with all its external fish-like attributes – shape, scales, and fins – was caught in 1938 in the nets of a fishing boat off the South African coast, in the depths of the Indian Ocean. Two meters long and hitherto unknown to science, its strange appearance surprised Marjorie Courtney-Latimer, curator at the East London Museum of South Africa and frequent visitor to her local port in search of new specimens to add to the museum's collections. Having made a summary description, she passed the specimen onto ichthyologist J. L. B. Smith, who identified it as a coelacanth and named it *Latimeria chalumnae* in her honor. It was the first living coelacanth discovered. Before this miraculous catch the animal was known only in the fossil record.

It was a shock and revelation for Smith, who, together with his wife, spent many of his remaining days implacably scouring the coast of the Indian Ocean in search of further specimens and offering rewards to local fishermen if they could haul another coelacanth up to the surface. The stubborn determination of the couple bore fruit 14 years later, when

1.2. The famous coelacanth. Its capture in 1938 has caused the spilling of much ink. Until then this animal was only known as a fossil, the youngest being 70 million years old! However, the coelacanth is not a "living fossil," just an organism that has traversed time without important modification to its external morphology.

The Great Transition 3

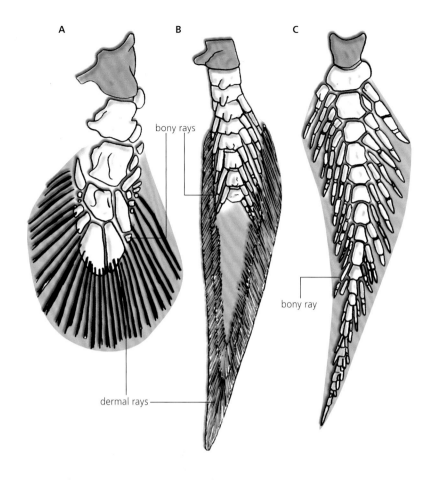

1.3. (A) the pectoral lobed fin of a coelacanth; (B) a tetrapodomorph; (C) a lungfish. Both lobed fin and monobasal articulation (in grey on the drawings, made up of a single bony head that attaches the fin to the scapular girdle) are characteristics of the sarcopterygians (some other vertebrates possess monobasal articulations). The sarcopterygians include ichthyans (of fishlike form) and the tetrapods (fig. 1.4).

a second specimen was caught off the coast of the Comoros. Despite Smith's tragic demise (he committed suicide), the famous ichthyologist's work remains inscribed in history. Although numerous fossil coelacanths had been discovered in sediments dating from the Devonian to the Late Cretaceous (410–370 Ma), none had yet been brought to light from the Cenozoic (an era combining two periods previously known as the Tertiary and Quaternary, extending from 65 million years ago to the present day). What is more, paleontologists thought, logically, that coelacanths died out at the same time as the non-avian dinosaurs, at the boundary between the Cretaceous and the Cenozoic. This is how the apparent disappearance of this peculiar fish was explained away, and no one imagined that a living coelacanth would ever be found.

Absence of proof, however, is not proof of absence – and, once again, history shows that theories put forward in paleontology, as in any other field of science, can be easily overturned in light of new discoveries. Plausible as it was, the traditional scenario of the disappearance of this curious blue fish did not survive the capture of the first living representative of the coelacanth group on the eve of World War II. The unexpected return of the coelacanth after 70 million years of absence became the talk of the town. The media jumped on the discovery – this "prehistoric survivor," "fish from the Dawn of Time," and "antediluvian ancestor" quickly made

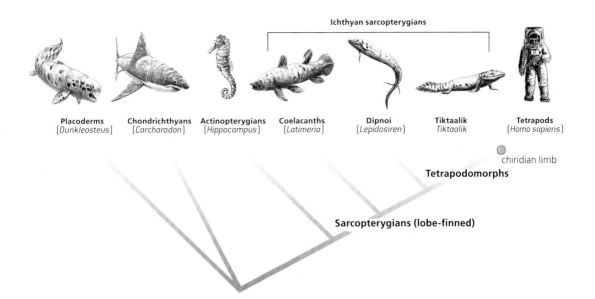

1.4. Relationships between the tetrapods and ichthyan forms: placoderms, chondrichthyans, ray-finned fish (actinopterygians), coelacanths, the lungfish and tetrapodomorphs. The coelacanths, lungfish and tetrapodomorphs are ichthyan sarcopterygians (fishlike). They are characterized notably by their lobed fins. Tetrapods are sarcopterygians that possess chiridian limbs. The ichthyan tetrapodomorphs are the ichthyan sarcopterygians most closely related to the tetrapods (species known only from the fossil record; see p. 9).

the front pages. The coelacanth attracted a plethora of names that were more or less incorrect, the most popular being that of "living fossil." From snippets in the popular press to specialist articles, from magazines on cryptozoology to treatises on comparative anatomy, a wealth of literature was spawned by this sensational discovery. However, the notion of the coelacanth being a "living fossil" is nonsense.

The poor coelacanth was saddled with this title for belonging to a supposedly fossil group, and yet it had been found alive and kicking! We cannot associate the terms "fossil" and "living" without doing harm to evolutionary concepts. We can consider someone "living" as being a "fossil" when the former displays an advanced state of aging. The only injustice done is to the person concerned. But if we juxtapose the terms "living" and "fossil" to qualify an entire taxon, we considerably alter the face of natural science; we would be constrained to classify most living species in the same category.

On close inspection, nearly all the life forms that inhabit our planet today have a long history – often longer than ours – and have morphologically similar "equivalents" in the fossil record. Cockroaches are known from strata dating back more than 300 million years. The same can be said for nearly all insects and arthropods. Bacteria – those that peacefully colonize our intestines, for example – are among the oldest forms of life on Earth and their evolutionary history stretches back more than 550 million years.

So why label the coelacanth a "living fossil," but not cockroaches or bacteria? This term, in fact, gives us a false idea of evolution. A so-called "living fossil" is not a carbon copy of an organism found in the fossil record: it does not live in the same environment as its ancestors, and its morphology is not strictly the same as theirs. Granted, there clearly are ancient species, like the coelacanth, that have traversed the ages, but scientists prefer to talk of panchronic species – not in the interest of political

The Great Transition   5

correctness, but because the notion of a panchronic taxon carries fewer preconceptions and is less emotive than that of "living fossil." Caution is the order of the day.

The celebrity of the coelacanth is due not only to its status as a panchronic species, but also to its special morphological traits. Apart from its intriguing blue color, it possesses a lung, albeit reduced and nonfunctional, and imposing fins firmly attached to its body. Lungs and mobile appendages are characters that evoke the tetrapods. The coelacanth's fin is lobed—that is to say, it is made up of bony rays and dermal rays (lepidotrichs). Some of these rays evoke the digits of tetrapods but, as we shall see next, this is just a similitude: the coelacanth is not a tetrapod. The lobed fin is attached to the scapular girdle by only one bony articulation, known as the monobasal articulation (fig. 1.3). Lobed fins and monobasal articulations (although present in much more ancient fish, such as the placoderms) are the distinguishing traits of the sarcopterygians (from the Greek "sarco" meaning flesh and "pterygium" meaning membrane), a group of vertebrates that includes lobe-finned fish such as the coelacanth (these are ichthyan sarcopterygians) and the tetrapods (fig. 1.4). The ichthyan sarcopterygians are therefore more closely related to the tetrapods than are other fish, such as those with rayed fins (the actinopterygians). This is why the group known as fish is in fact a grade, not a clade, and the name has no evolutionary meaning.

Sarcopterygians That Are Never Short of Breath: The Dipnoi

The title of living ichthyan sarcopterygian most closely related to the tetrapods does not go to the coelacanth, but to animals that are quite rare and almost as famous: the dipnoi. Also known as lungfish, these are vertebrates with an eel-like appearance (figs. 1.5 and 1.6). Six species exist today, divided into three genera (and into three families, for sticklers for Linnaean taxonomy), which can be found in their natural freshwater habitat in the Southern Hemisphere: *Lepidosiren* (*L. paradoxa*) in South America, *Protopterus* (four species) in Africa, and *Neoceratodus* (*N. forsteri*) in Australia. However, the dipnoi—appearing 380 million years ago during the Devonian and displaying much greater diversity than living examples—are better represented in the fossil record. Because of their peculiar lifestyle and their double respiration—both pulmonary and gilled (hence the name dipnoi, which means "of double respiration")—these sarcopterygians are curious in the extreme.

Discreet and seeming to defy the march of time, living dipnoi possess small eyes on the sides of the skull and a long body with delicate paired fins that resemble filaments (especially in *Lepidosiren* and *Protopterus*). The caudal fin starts at the center of the back and fuses with the anal fin. Although representatives of the genus *Lepidosiren* can attain a length of up to 1.25 meters, the record goes to *Neoceradotus*, at 1.6 meters and 40 kilograms! This Australian dipnoi, slightly different from the rest due to its large scales and greater corpulence, also holds the record for longevity: the Shedd Aquarium in Queensland, Australia, has a specimen over 80

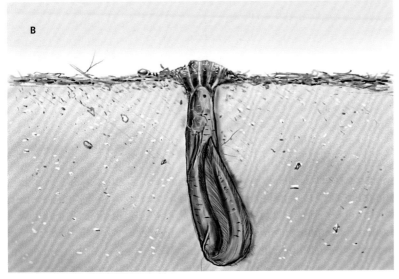

1.5. Scenes from a dipnoi's life (*Lepidosiren*). (A) Tied to an aquatic environment by its morphology, *Lepidosiren* has to return to the surface to breathe with its lungs. (B) These lungs are also used during the dry season, when it buries itself in a cocoon of mud and mucus and breathes air through small holes dug in its burrow.

years old and known as "Grandad" or "Pappy." This is the oldest dipnoi in captivity. Could this longevity be linked to the fact that this species, highly protected in Australia, grows at a slower pace than its South American and African cousins?

It is above all the functioning lungs of the living dipnoi that cause a sensation. Although the Australian species has only one lung and breathes mostly through its gills, *Lepidosiren* and *Protopterus* have two lungs that do most, if not all, of the adult individual's breathing. It goes without saying that lungs evoke images of the land-based tetrapod rather than of the fish in the deep blue sea. But the dipnoi are not tetrapods: they do not possess chiridian limbs.

The dipnoi's lung is relatively complex. It has alveoli (a character shared with the tetrapods) and opens from the same diverticle in the

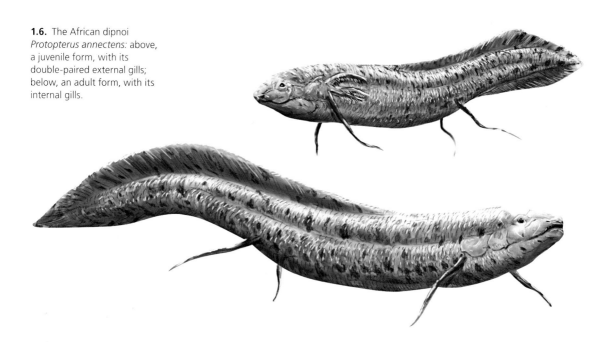

1.6. The African dipnoi *Protopterus annectens:* above, a juvenile form, with its double-paired external gills; below, an adult form, with its internal gills.

esophagus that forms the lungs of tetrapods and coelacanths. The lungs pump oxygen from the air, but when all is said and done, these organs are no more evolved than gills that pump oxygen from the water. There is no such thing as an organ more sophisticated than any other, just as there is no species more evolved than another. An organism equipped with gills is not inferior to one with lungs. Nevertheless, gills are too often associated with an inferior mode of living, a "base aquatic existence" verging on the unhealthy. William Alland's 1954 horror film, *Creature from the Black Lagoon*, perfectly illustrates our "gill-phobia," blending the mixture of fear and disgust that gills inspire in us and, consequently, anything equipped with them (fig. 1.7). The film depicts the cruelty of humans; the poor monster, with its enlarged external gills, has feelings and emotions as humans do but is, nevertheless, pitilessly hunted down.

Gills, however, are marvels of evolution; they are organs as complex as lungs. In certain amphibians the internal gills are maintained in the basicranial cavity by an extremely complex and branched network of small bony elements, whereas the external gills spread out into an elaborate fan, allowing the organism to extract oxygen from the most turbid of environments. The fact that we breathe air does not make us tetrapods superior beings.

Dipnoi use their lungs at the end of the wet season—when the watercourses where they live dry up, and they are forced to live on dry land. *Lepidosiren*, like its African cousin *Protopterus*, constructs a cocoon of mud and mucus beneath the ground and, using its lungs, breathes air through small holes in its burrow. It remains in a cataleptic state for several months before the end of the dry season. Then, when the rains come, it emerges from its lethargy and leaves its cocoon. The male then

1.7. The Creature from the Black Lagoon and his impressive gills. In *Creature from the Black Lagoon* (dir. William Alland, 1954), a monster takes a fancy to the voluptuous Julie Adams. The tragic destiny of this half-human, half-amphibian mutant perfectly illustrates human cruelty toward organisms with gills. His "cousin," the Man from Atlantis, is equipped with gills that are much less obvious.

assiduously builds an egg chamber, a sort of subaquatic burrow that it uses during the incubation period. This is where it looks after its offspring for at least seven weeks – until they are able to breathe air. During this period it oxygenates the water with its fins, as the young still must respire through their gills.

Even in a fully aquatic environment, some dipnoi cannot bypass using their lungs. A *Lepidosiren* held under the water will drown, despite the fact it has gills. Although tied to its aquatic environment by its morphology and its fins, *Lepidosiren* must surface to breathe – a terrible constraint for a member of the ichthyan clan!

Almost Tetrapod: The Tetrapodomorphs

We have just seen what living ichthyan sarcopterygians, the animals most closely related to the tetrapods, look like. However, this impoverished sample of living species is nothing compared to the past diversity of this group. Numerous fossils exist, in fact, that look like tetrapods and are their closest relatives (much closer than the coelacanth and the dipnoi) despite not yet being tetrapods: they used to go under the name of osteolepiforms. These fossil ichthyan species constitute, with the tetrapods, the clade called the tetrapodomorphs. The tetrapodomorphs share several characteristics, such as very particular paired fins (transformed into chiridian limbs in tetrapods) and a humerus with a convex head that articulates with the scapular girdle. The non-tetrapod (ichthyan) tetrapodomorphs that are the closest to tetrapods correspond to Devonian fossils. Some were contemporaries of the early tetrapods.

Lately, research into the origins of the tetrapods has intensified. As recently as several decades ago, the transition that led to the tetrapods was documented only in the fossil record by *Eusthenopteron* on the tetrapodomorph side and *Ichthyostega* on the tetrapod side. That was all. Today things have changed beyond recognition: although the number of paleontologists that work on this subject is very limited compared to those interested in dinosaurs and primates—more mediagenic subjects—increasing numbers of specialists worldwide are concentrating their energies on this fascinating area. What did the ancestors most closely related to tetrapods look like? We can begin to answer this question thanks to the active hunt for fossils in various parts of the world, mainly in Devonian layers.

Several fossil tetrapodomorphs have been found, including ancient tetrapods that show the line separating ichthyan sarcopterygians and tetrapod sarcopterygians is becoming increasingly fine. Happily, almost every year we discover ichthyan tetrapodomorphs more and more closely resembling tetrapods, and tetrapods more and more closely resembling ichthyan tetrapodomorphs. Even though these new fossils are mostly fragmentary, researchers are already filling the gaps in this transition period of capital importance: they know what the most derived ichthyan tetrapodomorph looked like, and are also discovering older and older tetrapods. The evidence proves that the ichthyan tetrapodomorphs most closely related to the tetrapods look very much like tetrapods but, in the place of limbs, had fins! Below are a few examples.

The Textbook Case of *Eusthenopteron*

Eusthenopteron

Age Late Devonian – 385 Ma
Location Miguasha, Québec
Size Up to 1.20 m
Features Dorsal fin, symmetrical caudal fin; endocranium known
Classification Ichthyan tetrapodomorph sarcopterygian

Among the ichthyan tetrapodomorphs, *Eusthenopteron foordi* remains one of the most complete fossils and therefore one of the most often cited. This genus, first comprehensively described only in the 1920s, despite initial description in 1881, is to fossil ichthyan sarcopterygians what the coelacanth is to living ichthyan sarcopterygians—a star of sorts. A textbook standard, this fossil of up to 1.2 meters comes from the Late Devonian cliffs (385 Ma) in Quebec's Miguasha National Park (fig. 1.8). Like all sarcopterygians worthy of the name, *Eusthenopteron* possesses lobed fins. But even though the caudal fin describes a symmetrical arc, which—it could be said—is a typical ichthyan trait, we will discover that its anatomy, starting with the arrangement of the cranial bones, places it closer to the tetrapods.

Sold in 1925 for the modest sum of $50 to the Natural History Museum of Stockholm by Joseph Landry, a Miguasha farmer, this fossil, which now figures on a Canadian stamp, owes its celebrity to Swedish paleontologists. It was eminent paleoichthyologist Erik Jarvik who carried out the first detailed study of this singular genus, starting with its cranial anatomy.

Well before the days of scanners and microtomography (see chapter 5), this scientist manually sectioned the skull of *Eusthenopteron* into slices, each one fractions of a millimeter thick, which he then remounted onto fine wax plates—an astonishingly painstaking job. These plates were then stuck back together to obtain a final model that was three-dimensional and magnified more than tenfold. This giant model beautifully illustrates the internal anatomy of the skull (fig. 1.9). The Swedish professor noticed that the endocranium (the internal structure of the skull) was capable of a double articulation. Known as the intracranial articulation, this was subsequently observed in the coelacanth and other fossil ichthyan sarcopterygians. Jarvik also noticed that the cranial structure of *Eusthenopteron* presented a mosaic of characteristics, some of which bore a curious resemblance to those of tetrapods known at that time. Imagine the surprise! This was most visible with the dermal bones of the skull roof, the pattern of which closely resembles that of tetrapods like *Ichthyostega* and *Acanthostega*.

This was not all: the dental enamel and dentine of *Eusthenopteron* was composed of tightly folded meandriform plates, much like the walls of a labyrinth. And yet these sinuous folds of enamel were found in

1.8. *Eusthenopteron* from the Late Devonian (385 million years), remains the most famous fossil ichthyan sarcopterygian as it was, for many years, considered the closest relative of the tetrapods. (A) Cast on display at the Swedish National Museum of Natural History, Stockholm. (B) Reconstitution of the animal in its coastal aquatic habitat.

Photo: Sébastien Steyer (CNRS/Muséum National d'Histoire Naturelle, Paris).

1.9. The wax model of the skull of *Eusthenopteron*, realized in the last century by Professor Erik Jarvik. This specimen, worthy of Madame Tussaud, comes from the collections of the Swedish National Museum of Natural History.

Photo: Sébastien Steyer (CNRS/Muséum National d'Histoire Naturelle, Paris).

1.10. The bones of the pectoral and pelvic fins of *Eusthenopteron*. They are homologous to those of the chiridian limbs of tetrapods – with the exception of the lepidotrichs, which are not homologous to digits (see fig. 1.11). *Eusthenopteron* is, therefore, an ichthyan tetrapodomorph, not a tetrapod. At right is a cast of the skeleton of the pectoral fin.

From Carl Buell, olduvai-george.com. Photo: Sébastien Steyer.

ancient tetrapods, which were, therefore, called labyrinthodonts (literally, with teeth of a labyrinthine structure). As such a structure was also present in ichthyan sarcopterygians, ancient tetrapods could no longer be qualified in this manner.

Erik Jarvik did not stop at the skull of *Eusthenopteron*. He also addressed his energies to the rest of the skeleton of this strange sarcopterygian and noticed that its fins (pectoral and pelvic, fig. 1.10) possessed bony structures homologous to those found in tetrapod limbs (fig. 1.11). The similarity was so striking that the first reconstructions of *Eusthenopteron* showed it as a form leaving the water and loitering on the shoreline (let us not forget that, at that time, the chiridian limb was associated with a

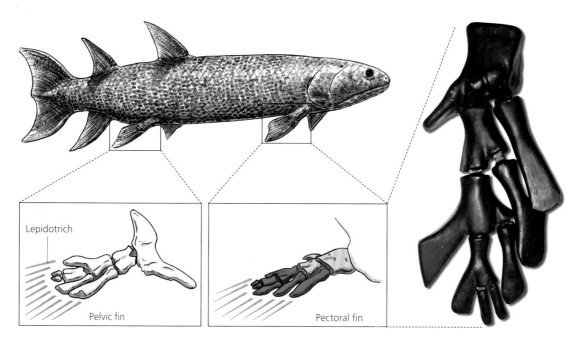

12   Earth before the Dinosaurs

terrestrial way of life). Today, however, the majority of paleontologists agree on its having been marine and pelagic (living in open waters).

Due to its mosaic of characters—both tetrapod and ichthyan tetrapodomorph—coupled with an exceptional state of preservation, *Eusthenopteron* is unseatable as a textbook classic and still has much to teach us.

The "Pandora Fish": *Panderichthys*

Another ichthyan tetrapodomorph worthy of our attention is *Panderichthys*, discovered in the 1940s in the Devonian deposits of Latvia and Russia (370 to 360 Ma). This sarcopterygian opens a new window on evolution and opens up a veritable Pandora's box. Although less complete than *Eusthenopteron*, it is closer to the tetrapods from a phylogenetic point of view and is a source of wonderment to specialists.

Despite our as yet partial knowledge of its anatomy, this peculiar sarcopterygian is nearly all tetrapod (figs. 1.12 and 1.13). Quite large (90 cm to 1.30 m), it apparently lacks a dorsal fin but, unlike its cousin *Eusthenopteron*, could have been equipped with an elongated caudal fin. This last characteristic, if proven, links it even more closely to the tetrapods. *Panderichthys* also has a very flat, triangular skull with dorsal orbits, as do its tetrapod cousins *Ichthyostega* and *Acanthostega*. Like a tetrapod, its intracranial articulation (the movable joint inside the skull that characterizes the sarcopterygians) is no longer externally visible. Among other things, *Panderichthys* is furnished with a spiracle, a natural opening behind the eyes that elongates into a kind of snorkel, allowing the animal to breathe water when lying buried in muddy sediments. This spiracle transformed, in tetrapods, into one of the bones that make up the inner ear. Finally—and altogether remarkable—its humerus is longer than that of most ichthyan sarcopterygians and its rigid vertebral column is much more suggestive of a tetrapod than a fish. Nevertheless, its pectoral fins are still furnished with dermal rays (lepidotrichs) and not with digits: it is an ichthyan tetrapodomorph, not a tetrapod. *Panderichthys* also has a

Panderichthys

Age Middle to Late Devonian—380 Ma
Location Latvia and Russia
Size Up to 1.30 m
Features Flat, triangular skull; snout; orbits in dorsal position
Classification Ichthyan tetrapodomorph sarcopterygian

1.11. Pectoral fin skeleton of *Eusthenopteron, Panderichthys,* and *Tiktaalik;* and the forelimb of tetrapods *Acanthostega, Ichthyostega,* and *Tulerpeton.* All the pectoral fin bones of the ichthyan sarcopterygians are homologous to those of the tetrapod limbs, with the exception of the lepidotrichs, which do not correspond to modified digits (see chapter 2).

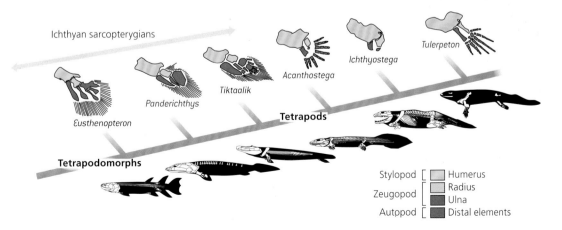

kind of snout, but this attribute, as we shall soon discover, is not exclusive to the tetrapods.

Two species of *Panderichthys* coexisted during the Middle and Late Devonian: *Panderichthys stolbovi*, known only from cranial and mandibular fragments, and the more complete *Panderichthys rhombolepis*. Analysis of sediments yielding these fossils suggests that *Panderichthys* could

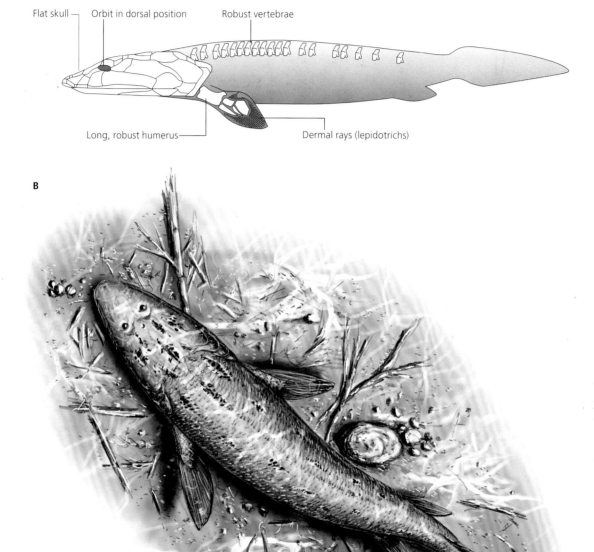

1.12. *Panderichthys*, a tetrapodomorph from the Devonian of Latvia and Russia (370–360 Ma): (A) preserved skeletal elements and silhouette; (B) reconstruction of the animal in shallow waters. *Panderichthys* displays a surprising mosaic of ichthyan (fins) and tetrapod (some skull elements) characters.

1.13. A specimen of *Panderichthys*, dating from the end of the Middle Devonian, discovered in Latvia in 1972. This fossil has been painstakingly prepared from its clay matrix, the nature of which accounts for its poor state of preservation.

Photo reproduced with the kind permission of Elga Mark-Kurik (Tallinn Technical University, Geological Institute, Estonia).

well have ventured into coastal and estuarine environments, which are rather brackish in nature. Moreover, its flattened skull with raised orbits, together with an elongated body, is evocative of an aquatic and benthic lifestyle. *Panderichthys*, therefore, probably inhabited shallow waters.

A colleague from Québec, Catherine Boisvert, is currently working in Uppsala, Sweden. Upon closer inspection of *Panderichthys*, she noticed anatomical details that suggest this organism could hang around in muddy sediments and wriggle from one pond to another, much like some catfish do today. Boisvert's scenario rekindles debate as to the degree of "terrestriality" in ichthyan tetrapodomorphs: like the famous Devonian tetrapods, *Ichthyostega* and *Acanthostega*, which have been reinterpreted as aquatic (we shall discover why), some ichthyan tetrapodomorphs already could have ventured onto dry land! "It's all upside down!" clamor certain colleagues, in a state of denial that these fins could be used on dry land. After all, given the new interpretations of *Ichthyostega* and *Acanthostega*, shouldn't the first (chiridian) limbs now be associated with swimming?

Tetrapod Yesterday, Ichthyan Tetrapodomorph Today: *Elpistostege*

Like its cousin *Eusthenopteron*, *Elpistostege* belongs to the paleofauna of the Miguasha National Park of Québec. It dates from the Late Devonian (385 Ma). The first specimen was discovered in 1938: a fragment of a flat skull with a long snout that, due to these characteristics, was identified by the paleontologist Stanley Westoll as belonging to a tetrapod. But since the 1980s, new specimens, all from Miguasha, have changed matters: certain skull fragments were found in association with vertebrae and scales that looked very much like those of *Panderichthys*. This aroused the curiosity of Hans-Peter Schultze, then of the Museum für Naturkunde, Berlin, and Marius Arsenault of the Miguasha National Park: both decided to re-examine the animal, this time in the minutest detail. It was after comparing material from old collections with this recent discovery that my colleagues first had doubts about the interpretation of this strange tetrapod, which was, in fact, nothing other than an ichthyan tetrapodomorph and quite closely related to *Panderichthys*. What was it that had led Stanley Westoll astray? It was that, unlike its distant cousin *Panderichthys*, *Elpistostege* displays orbits which are proportionately smaller and

Elpistostege

Age Late Devonian–380 Ma
Location Miguasha, Québec
Size Metric
Features Long snout; small round orbits in dorsal position
Classification Ichthyan tetrapodomorph sarcopterygian

The Great Transition   15

1.14. *Elpistostege* (380 Ma): (A) preserved skeletal elements and silhouette; (B) reconstruction of the animal ready to lunge at the first prey that comes within reach. This ichthyan tetrapodomorph – less complete than its cousin *Panderichthys* – had a very flat skull and a long snout. These traits evoke the early tetrapods to such a degree that *Elpistostege* used to be classified as one.

rounder, but also and above all, has a much longer snout (figs. 1.14 and 1.15): another feature thought to be exclusive to the tetrapods that spills over into the tetrapodomorphs.

It is not the first time that a fossil has changed identity; nor will it be the last. Science in general (and paleontology in particular) furnishes us with numerous examples. It shows that a fossil species can change position in the tree of life in light of new discoveries, especially if first described from fragmentary remains. In the 1980s an ancient tetrapod – *Elpistostege* – was rebaptized as an ichthyan tetrapodomorph. Today we can expect a reversal in the trend: when we rummage through the dusty drawers of European museums, those bony fragments from the Late Devonian previously attributed to ichthyan sarcopterygians or simply unidentified could possibly turn out to be tetrapods – on the condition that the dividing line between ichthyan tetrapodomorphs and tetrapods becomes increasingly vague. This is because nature itself cannot be pigeonholed. There

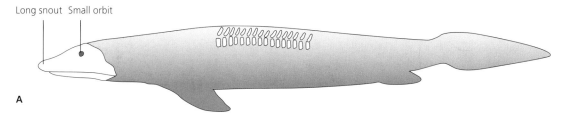

1.15. The skull of *Elpistostege* in dorsal view (part of the postorbital region has not been preserved).

Photo reproduced with the kind permission of the Miguasha National Park (Québec).

Orbits

are no pre-existing groups (taxa): it is humans that, by intellectual artifice, define and modify them in the light of our expanding knowledge.

In what environment did *Elpistostege* live? The Devonian ichthyofauna of Miguasha was long considered to be lake dwelling. Today, based on new sedimentological analyses, it is considered to have been more estuarine. It is therefore probable that *Elpistostege*, like its cousin *Eusthenopteron* (also from Miguasha), was adapted to brackish rather than freshwater.

The Sinking of the *Tiktaalik*

It was very recently, in 2006, that North American colleagues published, in the scientific journal *Nature*, a description of a new sarcopterygian fossil – proof that research and new finds are constantly emerging on the beginnings of the tetrapods. The publication concerns a tetrapodomorph that outdoes all others in its exceptional state of preservation: several dozen fossils were unearthed from the Late Devonian (Frasnian), in the Nunavut Territory of the Canadian Arctic. Of the numerous specimens from the frozen North, three had skulls, and articulated scapular girdles with pectoral fins in connection: a gold mine for paleontologists! This new taxon bears the strange name of *Tiktaalik* – "big river fish" in an Inuit

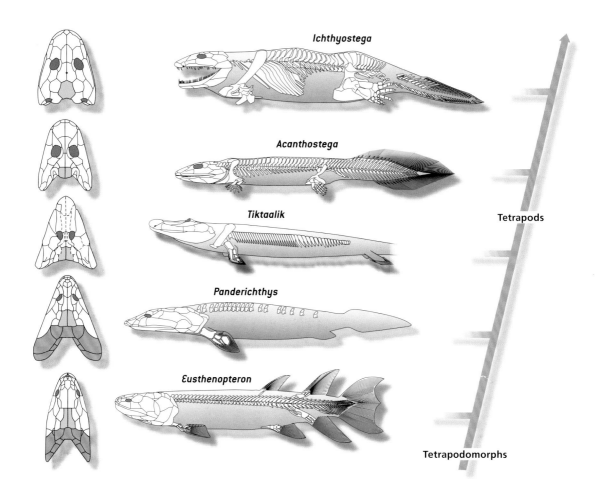

1.16. The relationships between *Eusthenopteron*, *Panderichthys*, *Tiktaalik*, *Acanthostega*, and *Ichthyostega*. On *Tiktaalik's* skull, as in the other early tetrapods, the opercular bones (in purple) have disappeared and the postparietals are reduced. *Tiktaalik* has nothing ichthyan about it except its fins: it is the closest known relative of the tetrapods.

dialect. This ichthyan sarcopterygian, which resembles a crocodile, is around 375 million years old.

Better preserved than the others, *Tiktaalik* upstages its cousins *Panderichthys* and *Elpistostege* because it is now considered to be the ichthyan sarcopterygian tetrapodomorph most closely related to the tetrapods (fig. 1.16). It is hardly an exaggeration to say that the only thing ichthyan about *Tiktaalik* are its fins. Put another way, the lack of digits is the only thing keeping it from being a tetrapod. In the same vein, its skull roof is full of surprises: even if all the sutures have not been identified, the skull roof is almost identical to that of a Devonian tetrapod. The opercular bones (those that cover the gills) have disappeared and the postparietals (bones at the back of the skull) are reduced. This throws much into doubt: these characters, clearly present in *Tiktaalik*, can no longer be considered attributes exclusive to the tetrapods. *Tiktaalik*, in addition, has a long snout—much longer even than that of *Elpistostege*—and its reduced orbits are set close together on the skull roof, along with rather large otic notches at the back of the skull and a neck that appears very mobile—just like a tetrapod's.

As we have seen, the hypothesis had already been advanced that certain tetrapodomorphs, such as *Panderichthys*, could venture onto dry

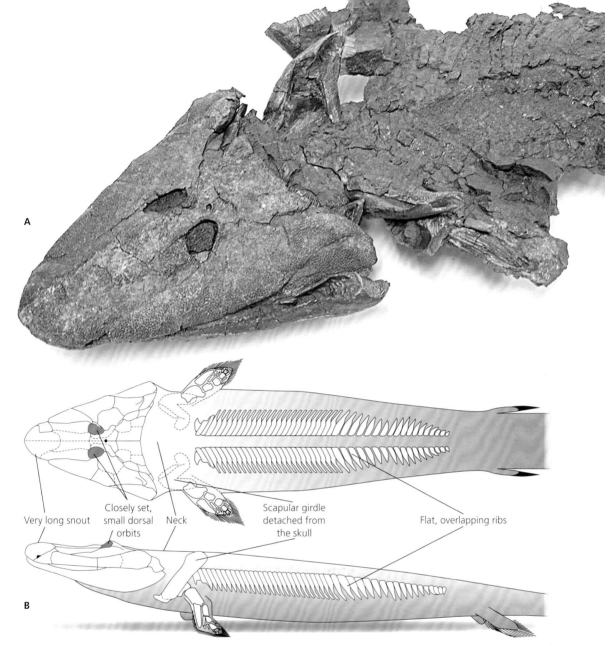

land. But this hypothesis lacked proof and, above all, complete fossils. *Tiktaalik* has changed all that. The fins of this extraordinary ichthyan sarcopterygian are unique and without any real equivalent in the living world. They possess dermal rays forming a veritable fan around the limbs (from front to back) and are supported by very robust bones resembling digits. The robustness of these fins may have permitted the animal to support its own weight out of the water. If we compare *Tiktaalik*'s fin with *Acanthostega*'s limb, the similarity is striking (see fig. 1.11): in both cases the humerus is robust and the ulna quite thickset. If we were to virtually stretch *Acanthostega*'s limb and replace the digits with dermal fins, the transformation would be complete: you would have in front of you the fin of *Tiktaalik*!

1.17. *Tiktaalik* (375 Ma): (A) the fossil extracted from its matrix in an excellent state of preservation (three-quarter view); (B) reconstitution of the skeleton. *Tiktaalik*, the latest addition to the tetrapodomorph family, beats all the records: better preserved and perhaps more terrestrial than the others (see fig. 1.18), it is also the closest known relative of the tetrapods!

Photo reproduced with the kind permission of Ted Daeschler (Academy of Natural Sciences, Philadelphia).

The Great Transition   19

1.18. The march of *Tiktaalik*. Whether lurking in the shallows or upon the shore, the fins of this sarcopterygian must have acted as temporary supports. *Tiktaalik* could permit itself the odd sortie onto dry land to taste, as shown here, some nice plump arthropods. The vertebrates did not wait for chiridian limbs to emerge from the water!

Tiktaalik may have inhabited shallow muddy watercourses or swamps. We think the fossiliferous locality that yielded the specimens corresponds to an equatorial delta choked with lush vegetation (such as tree ferns). This is perhaps why *Tiktaalik* periodically came onto dry land, where it crawled around using its robust fins (figs. 1.17 and 1.18).

Tiktaalik is, therefore, a highly unusual ichthyan tetrapodomorph whose story is by no means over and done with. I would even be willing to bet that this fossil, given its exceptional preservation and close relation to the tetrapods (there is no ichthyan form closer known at present) will soon replace *Eusthenopteron* as a textbook standard. *Tiktaalik* has a long future ahead of it.

Unfortunately, although only recently exhumed, *Tiktaalik* is already being misinterpreted. Categorized and squeezed into a narrow gap between the ichthyan tetrapodomorphs and the tetrapods, the poor *Tiktaalik* is portrayed as "the missing link between fish and tetrapods"; a misinterpretation which would oblige us to place *Tiktaalik* in between two taxa and not on its own branch in the tree of life. Paleontologists, mindful of the relationships between the organisms they study, have to drive this point home time and time again: a fossil is not a "missing link" between an ancient species and one more recent, and paleontology is not the art of looking for "missing links," as is all too often said! Why? Simply because this notion of "missing link," like the term "living fossil" (see p. 5), has no place in the natural sciences. An organism should be seen as a biological entity that has a combination of characteristics (morphological, for example) and not something at an intermediate stage of evolution. To talk of "missing links" is to wrongly interpret evolution as a straight line from bacteria to humans, or a slow and gradual succession of organisms towards increasing "complexity"; on the contrary, evolution radiates in all directions and is random: it has neither trend nor direction. Worse still, sometimes talk of "missing links" masks or censors the reality of evolution: in their so-called scientific publications and decorous speeches, creationists also use the expression "missing link," but in this case to illustrate the different "stages of creation."

Tiktaalik is, therefore, not the "missing link" between fish and tetrapods. It simply possesses a mosaic of morphological characteristics that link it closely to the tetrapods. For the abovementioned reasons, it is surprising that, even in the prestigious scientific journal *Nature*, certain colleagues have described *Tiktaalik* using the ambiguous terms "transitional" or "intermediate." It is true that the morphology of this fossil can be considered as "intermediate" (owing to its interesting combination of characteristics), but to consider *Tiktaalik* itself as a transitional form is misleading. Why did the authors employ these terms? Did they cave in to editorial pressure? Were they aware that creationists, who also read *Nature*, could use this language for their own ends? Stephen Jay Gould, a fervent defender of evolution, fought against the scourge of creationism, condemning the sophistry of those whose methods consist of, among other things, taking words out of context and twisting them around to mean their opposite. Let us not shoot ourselves in the foot; we should use precise terminology that does not communicate false ideas of evolution – even if it is less bankable.

Tiktaalik

Age Late Devonian–375 Ma
Location Canadian Arctic
Size Up to 2 m
Features Long snout; separated skull and scapular girdle
Classification Sister taxon of all tetrapods

Anything New on the Tetrapod Front?

To sum up, in the Late Devonian, around 370 million years ago, many sarcopterygians "played at being tetrapods," if you will pardon the expression. This suggests, as we have already seen, an ichthyan tetrapodomorph–tetrapod boundary that is increasingly ill defined. But does this necessarily imply that ichthyan sarcopterygians made a gentle transition into tetrapods? It is true that the cladograms we have seen so far suggest a succession of morphological innovations at each step. This does not, however, necessarily imply that the species described emerged gradually (slowly and progressively). In fact, there is one clue that runs counter to the idea of a slow transformation: the digits. These appeared suddenly in the course of evolution with the tetrapods, and genetic development shows us that these structures can form rapidly in embryogenesis (see chapter 2).

However that may be, recent discoveries of new ichthyan tetrapodomorphs and the reinterpretation of certain fossils have somewhat dispelled the tetrapod "myth": several traits previously considered to be exclusive to tetrapods have now proved to be red herrings. Originally tetrapodomorph in nature, they have switched over to the ichthyan side of the tree of life. Let us summarize.

Neither the alveolated lung present in ichthyan sarcopterygians nor the monobasal articulation found even outside the sarcopterygian clade are the patented domain of the tetrapod. The same goes for the choana, the internal nostril that links the throat with the palate: this is also present in ichthyan sarcopterygians (such as *Eusthenopteron*) and, therefore, existed several million years before the appearance of limbs!

More recently still, after the redescription of *Elpistostege* and the discovery of *Tiktaalik*, the tetrapods have lost their "monopoly" on long snouts (fig. 1.19). And yet it was believed that the appearance of a long snout corresponded to the transition from an aspiration, or suction, mode of feeding (in aquatic vertebrates) to a prehensile mode (in terrestrial vertebrates which, we suppose, were the early tetrapods). But this was not the case. *Tiktaalik* could easily, thanks to the axial muscles situated at the back of the skull, raise its head to trap and swallow its prey as, perhaps, did the tetrapods *Acanthostega* and *Ichthyostega* and as living alligators still do today. Some ichthyan sarcopterygians, therefore, were capable of feeding in the same manner as living tetrapods! Once again, thanks to the discovery of *Tiktaalik*, the absence of opercular bones and reduced postparietals are no longer tetrapod characteristics.

Another tetrapod attribute also may soon defect to the tetrapodomorph camp: the long caudal fin, stretched and pointed posteriorly, which can be seen in *Acanthostega* and other basal tetrapods. Most ichthyan species possess a caudal fin that is posteriorly enlarged, as is the case in *Eusthenopteron* (the only fossil ichthyan tetrapodomorph whose tail is preserved). What did the caudal fins of *Elpistostege*, *Panderichthys*, and *Tiktaalik* look like? Fish or tetrapod tails? The habitat (rather benthic, perhaps even terrestrial), and the flat body shape of these sarcopterygians would suggest a tetrapod-like caudal fin! We do not yet have enough concrete paleontological evidence to favor this hypothesis, but work is in progress.

Before the discovery of *Tiktaalik*, paleontologists thought that all ichthyan forms were no-necks (whose pectoral girdle attached to the back of the skull) and that cervical vertebrae (generally isolating the head from the trunk) were tetrapod attributes. This is no longer the case: the skull of *Tiktaalik* is well separated from its scapular girdle. This is a big surprise! The isolation of the head from the trunk—a profound reorganization of the axial skeleton—is associated with a modification in the position of the muscles surrounding the head, jaws, and, above all, the back skull. This "deconsolidation" of the skull from the rest of the body appeared in the ichthyan tetrapodomorphs and, therefore, no longer remains a tetrapod characteristic. It must have contributed to the flexibility of the body.

With attributes disappearing at this rate, what will the tetrapods have left to call their own? Among the innovations associated with tetrapods there is, of course, the chiridian limb and its digits. But other characters clearly define the appearance of this new group; here are some of the principal ones.

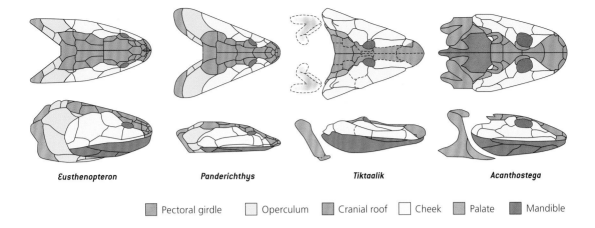

1.19. Comparison between skulls and pectoral girdles of *Eusthenopteron, Panderichthys, Tiktaalik,* and *Acanthostega*. Relative snout proportion is 38% in *Panderichthys*, 62% in *Tiktaalik*, and 55% in *Acanthostega*. An elongated snout is no longer a character exclusive to the tetrapods: it is present in *Tiktaalik*, which is a tetrapodomorph. *Tiktaalik* and *Acanthostega* (a tetrapod with a shorter snout than *Tiktaalik*) would have captured their prey in the same way as present-day alligators.

Courtesy of Dennis C. Murphy, www.devoniantimes.org.

In the cranial skeleton, certain bones are fused together at the front of the snout (perhaps lightening the skull) and in the jaw (maybe reinforcing this region). The lightening of the skull would have allowed greater mobility of the head, independent of the rest of the body, and strengthening the jaw would allow ventilation by a "buccal pump" in early tetrapods.

Dentition also differs greatly between the early tetrapods and the ichthyan tetrapodomorphs: in the latter, the jaw is equipped with an external row of small teeth (except at its extremity) and an internal row of fangs (on the coronoids). In Devonian tetrapods the opposite is true—the jaw is lined with an external row of large teeth (curving posteriorly) and an internal row of small teeth. These different types of dentition, for gripping in the first case (ichthyan) and for tearing in the second (tetrapod), suggest that these contemporary carnivorous organisms would have preyed on different animals, perhaps allowing the two groups to coexist for a time in the same littoral environments.

Tetrapod ears are also very peculiar (fig. 1.20). Our ichthyan cousins, apart from the lateral system on the bone or skin surface (this is the line you cut along to fillet your favorite fish) also possess an inner ear, a complex membranous system located behind the eye called the labyrinth. This is linked to the brain and is present in all vertebrates. In ichthyan forms it contains small calcareous elements known as otoliths (sometimes also called statoliths). These lie on a carpet of ciliated cells and communicate changes in pressure (sound) but, above all, they allow the animal to orient itself in the water. In tetrapods (with the exception of *Ichthyostega*) there are always otoliths in the ear, but the transmission of sound waves in the skull is amplified by a bony rod, the columella (or stirrup, in mammals), which serves as a piston that pushes against the tympanum. It is not an entirely new invention and is, in fact, derived from a bone our ichthyan ancestors possessed—the hyomandibular. This helped to support the jawbone and aided in the articulation of the gill cavity. The transformation of the hyomandibular into the columella in most tetrapods could be linked to the reduction of the back skull and modification of the jaw. The columella, however, would not be present

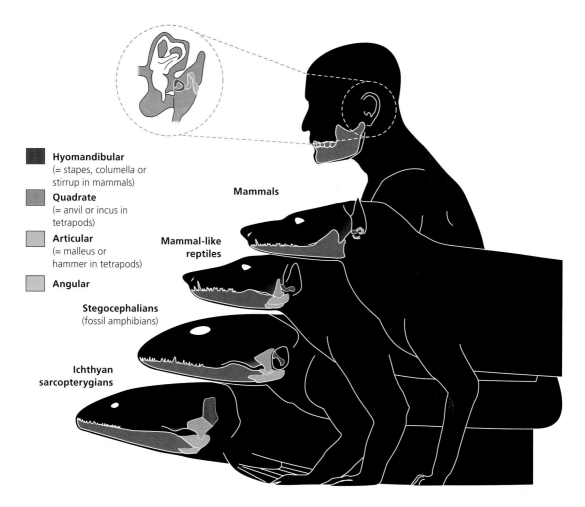

1.20. The mandible and inner ear of ichthyan sarcopterygians and tetrapods (stegocephalians, "mammal-like reptiles," and mammals). Tetrapod ears are very peculiar: inside, the transmission of sound waves is amplified by a bony rod, the columella (or stirrup, in mammals), which is homologous to the hyomandibular in the ichthyan sarcopterygian jaw.

in all tetrapods. *Ichthyostega*'s ear still leaves many researchers speechless (and often deaf). According to recent studies by my colleague Jennifer Clack of Cambridge University, it had a spoon-shaped hyomandibular. This probably vibrated in the hollow of its ear, "prefiguring," to a degree, the columella of its limbed cousins.

If we look now at the posterior anatomy of the tetrapods, we see the pelvis (basin) is elongated and joined to the vertebral column, forming a veritable girdle called the pelvic girdle (fig. 1.21). This, like the pectoral girdle, plays a role in locomotion by linking the trunk with the limbs. Like the digits, the pelvic girdle seems to be a specificity of the tetrapods that other sarcopterygians apparently do not share—or do not appear to. Indeed, observations of living tetrapod embryos show that the pectoral girdle can appear before the pelvic girdle and, therefore, the two do not necessarily form in a synchronized manner; was their evolution marked by an identical chronology? Can we suppose there was a "hybrid" animal with a robust pectoral girdle and chiridian forelimbs (as in *Ichthyostega*) but with posteriorly reduced girdle and fins? This fossil (chimera even!), half tetrapod, half ichthyan, would itself be a revolution in paleontology and a serious headache to classify.

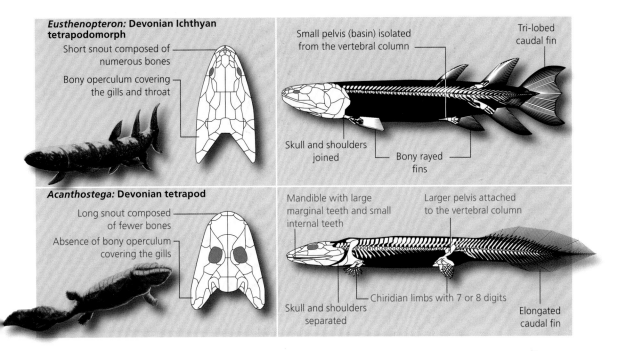

1.21. Comparison between an ichthyan tetrapodomorph (*Eusthenopteron*) and one of the most ancient tetrapods (*Acanthostega*). Thanks to the discovery of new fossils, attributes specific to the tetrapods have melted away. The type of dentition, a large pelvis joined to the vertebral column, and chiridian limbs remain exclusive to the tetrapods (in red). Characters in blue, although absent in *Eusthenopteron,* have been observed in other ichthyan tetrapodomorphs.

Let us conclude with an interesting observation; we saw before that the specific characteristics "hijacked" from the tetrapods were linked by adaptation to the scenario of inevitable tetrapod emergence from the water. Indeed, up until very recently, we wanted to fit each morphological trait to a function. For example, the lungs, air-breathing organs par excellence, were unique to the early tetrapods ("because they were terrestrial" being the kneejerk assumption). Inconveniently, the tetrapods are not the only sarcopterygians to possess lungs! The same goes for monobasal articulations: this "tetrapod characteristic" was seen as an ideal structure that appeared to facilitate walking. Bad luck once again, for the monobasal articulation is also present in fish! The list of "adaptations to colonize the land" goes on and on because what we imagined "terrestrial," at that time, rhymed with "tetrapod."

Today this state of affairs has changed enormously and discoveries have, once again, won over against adaptationist lines of thinking: those "typically terrestrial" characteristics that defined the tetrapods, in fact, appeared in the water. We will now find out why paleontologists are convinced of this.

Swimming before Walking

What did the early tetrapods look like and where did they live? Even if there are still many elements missing about the origin and radiation (the increase in diversity) of early tetrapods, researchers now have in their hands very interesting clues that go some way toward answering these questions.

The most complete ancient tetrapods are *Ichthyostega* and *Acanthostega*, fossil "stars" around 370 million years old (late Famennian). They

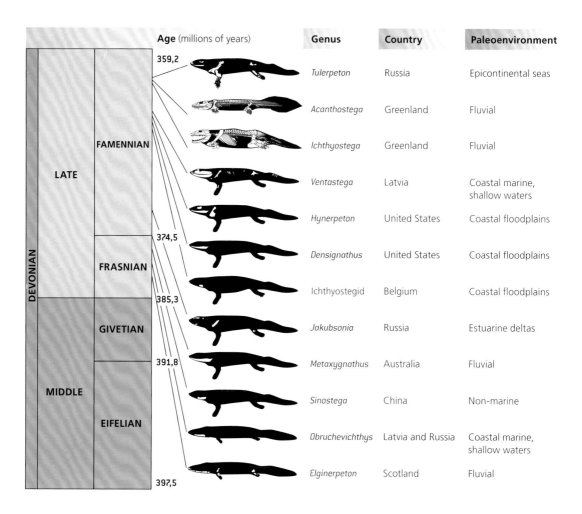

1.22. Early Devonian tetrapods: the state of play. *Ichthyostega* and *Acanthostega* are the "old" tetrapods (discovered in the 1930s) and are the best preserved. Since the mid-1990s, tetrapod discoveries in the Late Devonian have multiplied. Recently my colleagues have even found tetrapod trackways in the Middle Devonian of Poland. Will their skeletons also soon be found? In addition, most of the fossils have been discovered in sediments corresponding to coastal paleoenvironments.

were discovered in eastern Greenland and their descriptions were first published in the 1920s and 1930s by illustrious Swedish paleontologist Gunnar Säve-Söderbergh. It was amazing that, finally, the ancestors of all living tetrapods had been unearthed. *Ichthyostega* and *Acanthostega* held onto their title of oldest known tetrapods for more than half a century.

But all good things must come to an end, and since the 1990s we have witnessed an explosion of discoveries from around the globe. The oldest known tetrapods to date are from the middle Frasnian (380 Ma). They were, therefore, contemporaries of their ichthyan tetrapodomorph cousins. The number of tetrapod genera from the Late Devonian has consequently gone from 2 (*Ichthyostega* and *Acanthostega*) to more than 12 (figs. 1.22 and 1.23). To the list of "limbed stars" we must now add *Tulerpeton* (from the Tula region) and *Jakubsonia*, both from Russia; *Ventastega* from Latvia; *Obruchevichthys*, a tetrapod from Russia and Latvia bearing the misleading suffix of "ichthys"; *Hynerpeton* and *Densignathus* from the famous Red Hill site, Pennsylvania; *Elginerpeton* from Scotland; *Metaxygnathus* from New South Wales; *Sinostega* from the Ningxia Hui Autonomous Region in China; and a form close to *Ichthyostega* found in

26 *Earth before the Dinosaurs*

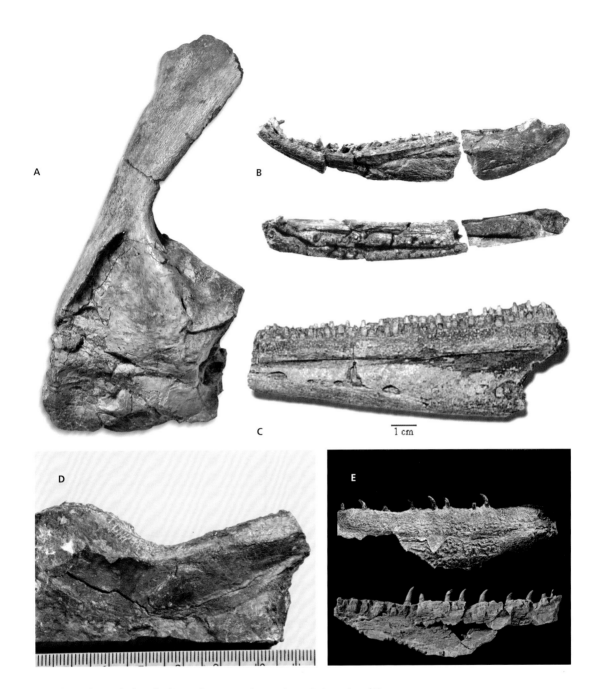

1.23. Several recently described Devonian tetrapods or, rather, what remains of them: (A) *Hynerpeton* and (B) *Densignathus,* United States; (C) *Elginerpeton,* Scotland; (D) *Sinostega,* China; and (E) an ichthyostegid from Belgium. These taxa are known only from their mandibles or their scapular girdles. Some elements clearly present tetrapod characteristics (like a certain tooth type that is posteriorly curved). Others, like the Chinese specimen, are more enigmatic and difficult to assign.

Photos reprinted with the kind permission of Ted Daeschler (Academy of Natural Sciences, Philadelphia; A and B); Per Ahlberg (Uppsala University, Sweden; C); Zhu Min (Institute of Vertebrate Paleontology and Paleoanthropology, Chinese Academy of Sciences, Beijing; D); and Gaël Clément (Muséum National d'Histoire Naturelle; E).

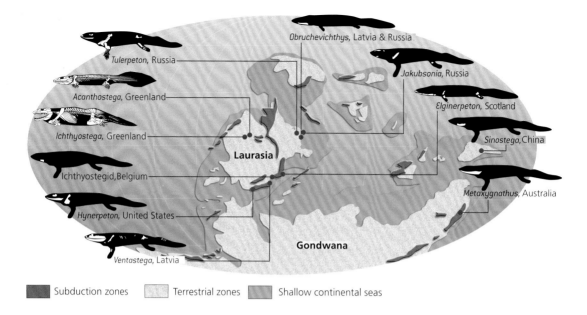

1.24. Worldwide distribution of Late Devonian tetrapods. We have indicated on this paleomap where the oldest fossil tetrapods to date have been found. Their extreme diversity and worldwide distribution suggest that the group is perhaps even older than the Late Devonian (see also fig. 1.25). This has just been confirmed by trackway discoveries from the Polish Middle Devonian.

Belgium by Gaël Clément's team from the Muséum National d'Histoire Naturelle in Paris. In addition, two or three new forms are discovered each year.

Older Origins?

Devonian tetrapods have grown in number in the last few years! The newcomers are not always as well preserved as their mentors *Ichthyostega* and *Acanthostega*. However, despite their fragmentary nature (known only from their mandibles and/or scapular girdles), comparisons can be made that show they were not all alike. Some of these tetrapods even display remarkable specializations; this suggests that they were much more numerous and diversified than previously thought. From the Late Devonian they seem so diverse and geographically dispersed (fig. 1.24) that we must ask ourselves if they really were the first—or do we need to look for other tetrapods in more ancient layers?

It could be that the tetrapods are older than we imagine and their oldest fossil representatives known today are the result of an evolutionary radiation that occurred before the Late Devonian. I would not be surprised to learn of the discovery of a completely new tetrapod from Middle Devonian layers. If this were to happen, then the story of the very early tetrapods would be a mystery between the Middle and Late Devonian, leaving a gaping hole in our knowledge of their evolution, as is already the case with Romer's Gap (named for twentieth-century American paleontologist Alfred Sherwood Romer) between the Early and Late Carboniferous (see chapter 3), and also Olson's Gap between the Early and Late Permian (see chapter 4). Should there be a Devonian gap, I would propose we call it "Janvier's Gap," in honor of French paleontologist

Philippe Janvier (Centre National de la Recherche Scientifique [CNRS] and Muséum National d'Histoire Naturelle, Paris), who studies early vertebrates. At the end of the day, whatever the gap—Janvier's, Romer's, or Olson's—is it not one of the duties of paleontology to fill it in?

We will feed this hypothesis of Janvier's Gap by mentioning strange fossil footprints, attributed to tetrapods that have been found in many parts of the globe (Australia, Brazil, Greenland, Scotland, and Ireland) in layers dating from the Late Devonian (Frasnian), and even the Middle Devonian (Givetian) (fig. 1.25). These curious trackways, apparently left on dry land, are doubly surprising in that Devonian tetrapods are considered to have mostly been aquatic forms! Although there is still animated debate over the precise age of these tracks and their real nature (some footprints could have been made in shallow water), they suggest that, in the Middle Devonian, tetrapods were already leaving traces behind them. Just after the French edition of this book was published, tetrapod trackways were discovered in the Middle Devonian of Poland. These well-identified and dated footprints have been found in an abandoned quarry in southeastern Poland, and they come from marine tidal flat sediments of early Middle Devonian (Eifelian) age. They are about 18 million years

1.25. Trackways in the Middle–Late Devonian of Ireland: Are these footprints left by a tetrapod? The question remains unanswered. It could be that the tetrapods are older than anticipated and that their oldest known fossil representatives were the result of an evolutionary radiation that occurred before the Late Devonian (see also fig. 1.24).

Photo reproduced with the kind permission of Christian Falipou.

older than the earliest tetrapod skeletal element. We now have proof that tetrapods were living before the Late Devonian, and that Janvier's Gap is a reality! We still need to find tetrapod body fossils (skeletal elements) in the Middle Devonian, of course, but once again, these discoveries perfectly illustrate that our knowledge of tetrapod evolution is, in turn, evolving.

The Mysterious Island: The Discovery of *Ichthyostega*

Let us return to the first discoveries of tetrapod fossils. In the 1920s and 1930s several geological surveys and cartographical expeditions were made by Swedish geologists and paleontologists to find out more about Greenland, an island then (and even today) shrouded in mystery, and whose desolate landscape and hostile environment continue to fascinate scientists. Naturalists collected case upon case of minerals, rocks, and fossils.

The paleontologist Säve-Söderbergh—who participated in some of these missions from 1931 to 1934, and again in 1936—began to separate from their rocky matrix the fossils he considered the most important in order to describe them. In doing so he left aside a caudal fin of a dipnoi that appeared more or less anodyne. Concentrating on other specimens, in 1932 he published a preliminary, but accurate, paper on a strange tetrapod. Mesmerized by the morphology of this animal, he arrived at the conclusion that this surprising fossil from the late Famennian (Latest Devonian—360 to 362 Ma) represented a crossroads between two worlds and decided to name it *Ichthyostega* (from the Greek "ichthyos" meaning fish, and "stegos" meaning plate—a term often reserved to describe ancient amphibians).

Ichthyostega

Age Late Devonian—362 to 360 Ma
Location Greenland
Size Up to 1.5 m
Features Rather high skull; robust girdles; hindlimbs with seven digits
Classification Basal tetrapod

After the premature death of Säve-Söderbergh in 1948, his colleagues, including Erik Jarvik (who also described *Eusthenopteron*; see above), interested themselves in the fin their departed friend had not had time to study. When they prepared it they got the surprise of their lives: it was not attached to the body of a dipnoi, but to the basin of a tetrapod with limbs—the same *Ichthyostega* previously described by Säve-Söderbergh! *Ichthyostega*, with its fishy tail, certainly deserved its name.

Indeed, these various Swedish paleontologists arrived at the same conclusion: *Ichthyostega*, from the Late Devonian of Greenland, was quite an exceptional organism, a strange (almost alien!) life form, the skeleton of which displays a mixture of ichthyan characters (such as gills and a caudal fin), and tetrapod characters (including legs and a skull with a reduced number of bones; fig. 1.26). It was believed that the "missing link" had just been found (yet again this name rears its ugly head!) "between fish and amphibian": the first "limbed fish," or the first "terrestrial vertebrate." This is how *Ichthyostega* was painted.

But how did such a "hybrid" come about? By what mystery of evolution could a fish "grow itself legs"? To answer these questions, an easy-to-digest

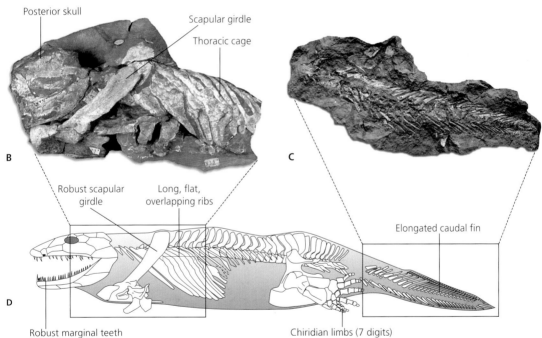

1.26. *Ichthyostega* from the Late Devonian (Late Famennian) of Greenland: (A) skull roof; (B) thoracic cage, posterior skull, and scapular girdle; (C) tail; (D) reconstitution of the skeleton and silhouette. Discovered in the 1930s, *Ichthyostega* is one of the most complete Devonian tetrapods.

Photos reproduced with the kind permission of Jennifer Clack (University Museum of Zoology, Cambridge); specimens belong to the Geological Museum of the University of Copenhagen, Denmark (A and B); and Gilles Cuny (Geological Museum of the University of Copenhagen; C). See also fig. 1.30 for hindlimb.

evolutionary scenario was invented; in it, we imagine sizeable carnivorous freshwater fish happily cohabiting a pond. Everything goes swimmingly for these organisms until one day, during a drought, the water level drops, forcing these fish to breathe air—which they can do thanks to their lungs—and then slither across muddy areas to find a deeper pond. This is how, slowly but surely over generations, fish developed limbs and transformed into *Ichthyostega*. This was touted as gospel truth from the

1.27. *Ichthyostega* – 360 Ma. Most recent analyses suggest that this tetrapod could probably have made sorties onto land, as seals do today, but would have spent most of its time in the water.

Acanthostega

Age Late Devonian – 360 Ma
Location Greenland
Size Up to 60 cm
Features Rather flat skull and body; polydactylous limbs
Classification Basal tetrapod

1920s to explain vertebrate conquest of the land and the emergence of the tetrapods.

Is this version of events still considered valid today? Definitely not! We shall see how findings over the last dozen years have shot down the myth of the conquest of the land. It is another prime example of how yesterday's theories can be quickly debunked by the fossils of today, just as the theories of today will be debunked by the fossils of tomorrow.

The End of a Myth: *Acanthostega* and Partners

As the first vertebrate with legs, *Ichthyostega* rapidly became a star of radio and newspapers, and even made an appearance in my school curriculum! In 1933 the description of another key fossil, also from the Latest Devonian (late Famennian) of Greenland, was published by Säve-Söderbergh and Jarvik. It was that of *Acanthostega* (figs. 1.28 and 1.29). *Acanthostega* also possessed chiridian limbs and, almost overnight, became *Ichthyostega*'s little cousin.

Expeditions followed in quick succession; their destination – the land of ice. *Ichthyostega* and *Acanthostega* had once lived there among the early tetrapods and were previously labeled terrestrial. Other specimens were found to fill out museum collections, right up until the 1970s; some of them lay forgotten, gathering dust. They were rediscovered in the 1990s, notably thanks to the work of Jennifer Clack.

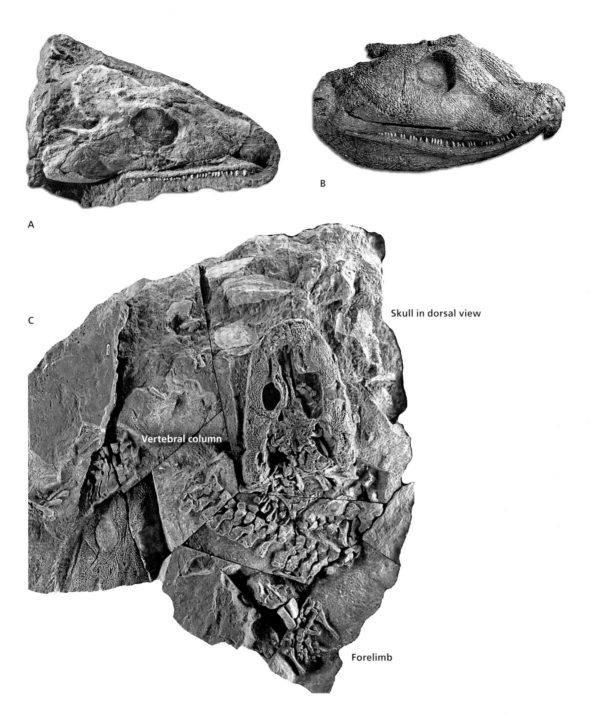

1.28. *Acanthostega* from the Late Devonian (Late Fammenian) of Greenland: (A) mid-skull in dorsolateral view; (B) complete skull in lateral right view (the snout is slightly "crushed"); (C) sub-complete skeleton. *Acanthostega* is the star of Devonian tetrapods. Reanalysis of several specimens by Jennifer Clack in the 1990s showed that this animal was rather aquatic, contrary to a commonly held belief dating back to the 1930s.

Photos reproduced with the kind permission of Gilles Cuny (Geological Museum of the University of Copenhagen, Denmark; A); Jennifer Clack (University Museum of Zoology, Cambridge; B and C); specimens belong to the Geological Museum of the University of Copenhagen, Denmark (B and C).

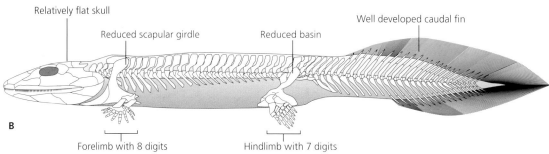

Relatively flat skull
Reduced scapular girdle
Reduced basin
Well developed caudal fin
Forelimb with 8 digits
Hindlimb with 7 digits

1.29. (A) *Acanthostega* in its element: the water. The moderately rigid limbs of this tetrapod must have been used as paddles and not for walking: it was apparently incapable of moving around on dry land. (B) Skeletal reconstruction and silhouette.

This is how *Acanthostega*, found in the same geological layers as *Ichthyostega*, is today recognized as the oldest and most complete tetrapod. Its detailed re-examination raised serious doubts about vertebrate conquest of the land: we now know that, first, these vertebrates did *not* abandon the water from a dried-up pond; second, they were *not* as terrestrial as once thought; and, third, ichthyan tetrapodomorph fins are not *completely* homologous to chiridian limbs (because a fin's dermal rays are not homologous to digits). How did we arrive at this conclusion?

Inside the walls of Cambridge's Museum of Zoology and Comparative Anatomy, preparator Sarah Finney, under the guidance of Jennifer Clack, painstakingly and with extreme caution prepared the Greenland fossils, some of which they had found themselves. Years of toil bore fruit and, as we now realize, thanks to these two colleagues, *Acanthostega*'s limbs were not used for walking. They were quite rigid and more paddle shaped than foot shaped. Moreover, *Acanthostega* possessed well-developed and apparently fully functioning gills. To top it off, its limbs did

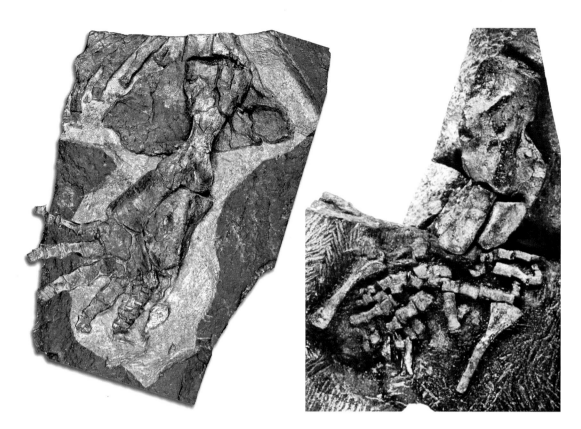

not have five digits but eight (fig. 1.30)! This truly put the cat among the pigeons, because everyone agreed that five digits was an archaic character going back to tetrapod origins. *Ichthyostega*'s hindlimbs possess seven digits; its forelimbs are not yet known. Other Devonian tetrapods have since been discovered around the world and some, like *Tulerpeton* (with six digits), have also proved to be polydactylous (fig. 1.31).

In the wake of Clack's work, the dusty drawers of Europe's natural history museums were reopened: mandibles previously attributed to ichthyan sarcopterygians were rigorously re-examined and some, surprisingly, showed tetrapod characters! Meanwhile, research on developmental genetics intensified, and it was discovered that the dermal rays of ichthyan fins were not homologous to tetrapod digits (some specialists talk of neoformations; see chapter 2). Today, biomechanical studies allow us to model the locomotion of these basal tetrapods and to propose new reconstitutions and it appears that (as Clack already proposed) *Acanthostega*'s articulations restricted its limbs' range of motion to such a degree that it was incapable of bringing them forward and thereby raising itself off the ground. They must, therefore, have served as paddles – confirming the aquatic lifestyle of this Devonian tetrapod!

Ichthyostega's forelimbs could not carry out flexion and extension, and its hindlimbs were flipper shaped and oriented towards the rear. What's more, *Ichthyostega*'s vertebral column did not allow lateral or undulatory movement (many modern reptiles and amphibians crawl

1.30. The hindlimb of *Ichthyostega* (left: seven digits) and the forelimb of *Acanthostega* (right: eight digits). In the living world, tetrapods have, except in cases of "accident," five digits at most. But this was not necessarily the case for their ancestors. *Photos reproduced with the kind permission of Jennifer Clack (University Museum of Zoology, Cambridge); specimens belong to the Geological Museum of the University of Copenhagen, Denmark.*

Tulerpeton

Age Late Devonian
Location Tula, Russia
Size Up to 1.5 m
Features Six-digit limbs; rather robust girdle and forelimbs
Classification Basal tetrapod

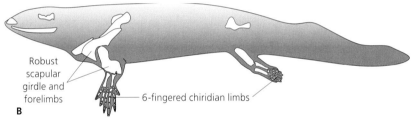

1.31. *Tulerpeton* from the Late Devonian of Russia: (A) Reconstitution; (B) Elements preserved and silhouette. This fossil is less complete than that of *Ichthyostega* or *Acanthostega*. Like them, it displays polydactyly (having more than five digits). Note that the quite robust scapular girdle suggests a more terrestrial lifestyle than that of its other Devonian cousins (despite the fact it was found in ancient estuarine sediments).

1.32. (below) Preserved elements of *Hynerpeton*'s skeleton, a tetrapod from the Late Devonian of the United States; so rare they can be counted on the digits of one hand!

1.33. (facing) *Hynerpeton* – 360 Ma. The Red Hill site in Pennsylvania has yielded numerous plant remains in association with those of *Hynerpeton*. It probably lived in a coastal environment very rich in plants and organic matter, as did other Devonian tetrapods.

sinuously along the ground). Limited vertical movements of the trunk could have allowed *Ichthyostega* to spend time on dry land, perhaps as seals do today (see fig. 1.27).

If the early tetrapods were not terrestrial, then what environment did they live in? Ideas on this have greatly changed since the first half of the twentieth century. The geologists of yesterday interpreted the sediments enclosing these fossils (mostly red sandstones) as continental deposits, essentially fluvial or lacustrine. Today, more detailed analyses suggest

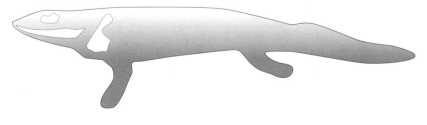

1.34. A fossil branch of *Archaeopteris,* considered the oldest tree in the world. It attained a height of up to 30 meters and belonged to the group of progymnosperms. These plants grew in humid environments and could have formed veritable coastal forests.

Photo © DK Limited/CORBIS

Hynerpeton

Age Late Devonian – 361 Ma
Location Red Hill, Pennsylvania
Size Metric
Features Rather flat mandible
Classification Basal tetrapod

coastal, estuarine, or lagoonal deposits. The early tetrapods could therefore tolerate brackish water and even seawater. The vast majority of living amphibians, a term that groups together all the non-amniotic tetrapods (see chapter 3), are freshwater animals, and cannot tolerate salt water. How did their fossil ancestors manage? Even if the question has yet to be answered in any detail, this euryhalinity (the ability to tolerate great variations in salinity) goes a long way toward explaining the wide distribution of the early tetrapods.

Whatever the answer, numerous other Devonian fossils have been unearthed in "tetrapod localities" throughout the world; a variety of plants, invertebrates and ichthyan groups including placoderms. All these make up the flora and fauna associated with the early tetrapods. These aquatic paleoenvironments were dense and rich in roots and plant debris (figs. 1.32 and 1.33). Reconstitutions evoke a mangrove-type environment in which plants called *Archaeopteris* (fig. 1.34; not to be confused with the feathered dinosaur *Archaeopteryx*) "prefigured" the mangrove of today. How could the basal tetrapods swim in these murky waters? What were the principal functions of their limbs? To answer these questions, we must first find out how limbs are formed. This is where developmental genetics becomes of the utmost importance.

Paleontological Enquiry and the Randomness of Fossilization

FACED WITH A FOSSIL, THE PALEONTOLOGIST FINDS HIMSELF in the role of a detective carrying out an investigation into the circumstances of a death as well as trying to determine identity. The "victim" in this case, however, died millions of years ago! A fossil is the remains of an organism. Not everything gets preserved, of course, and the environment in which it was deposited (or preserved in) is not necessarily the environment in which the organism lived. The depositional environment is deduced after sedimentological and paleoenvironmental analyses of the sedimentary matrix, whereas the living environment is obtained directly from the morphology of the skeleton (notably through biomechanical studies). There is not necessarily a link between these two environments, which complicates matters. Before eventual fossilization, some cadavers can, for example, float long distances, so there are many random factors that come into play between the death of an organism and the discovery of its fossil.

The study of burial and fossilization phenomena is known as taphonomy. Taphonomists, necromancers of the past, attempt to reconstitute the paleoenvironment in which a fossil is found. Together with paleontologists, sedimentologists, stratigraphists, geochemists, and paleoclimatologists, they analyze fossil assemblages: How did the fossils get here? We must not forget that a fossil site is not a true mirror of the biodiversity of a given period. It is more like an old family photo in which most of the figures have been effaced over time.

Let's take the example of *Tulerpeton* from Russia (see fig. 1.31) which, with its relatively robust scapular girdle and limbs, appears somewhat terrestrial. It was, however, discovered in rather marine sediments. Such "anomalies" are not unusual if we consider the type of coastal environments favored by the early tetrapods, but it does render matters more complicated. A simple walk on the beach today reveals the cadavers of marine organisms mixed with those of terrestrial organisms, and perhaps only a fraction will, given time, fossilize. What will future paleontologists make of this?

1.35. *Palaeomacropis eocenicus*, a bee trapped in amber from the Lower Eocene of L'Oise, France (50 Ma). Amber is probably the most exceptional medium of fossilization; in it, even the soft parts of imprisoned organisms are "frozen" in time.

Photo reproduced with the kind permission of André "Dédé" Nel (Muséum National d'Histoire Naturelle).

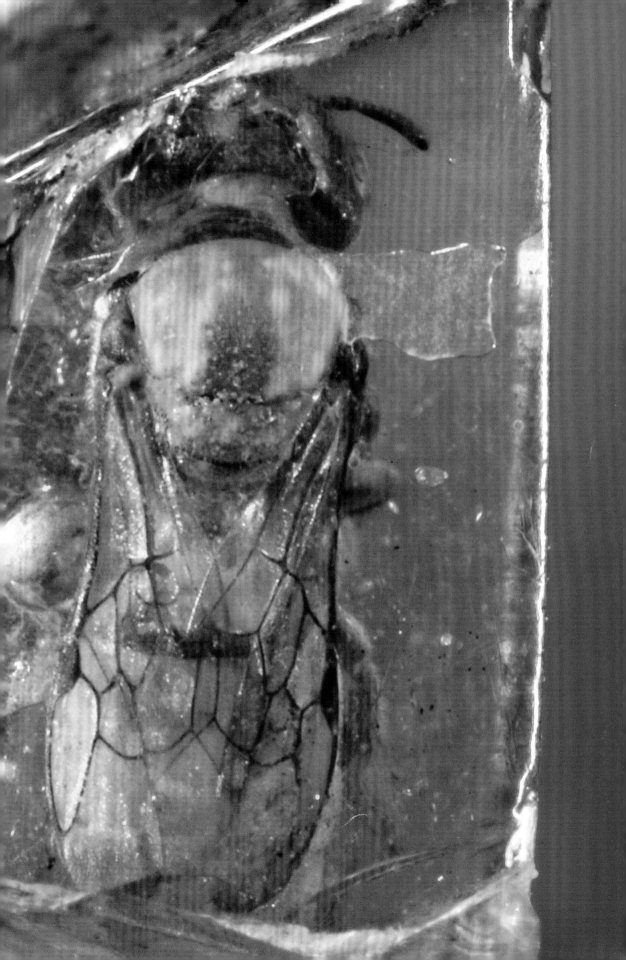

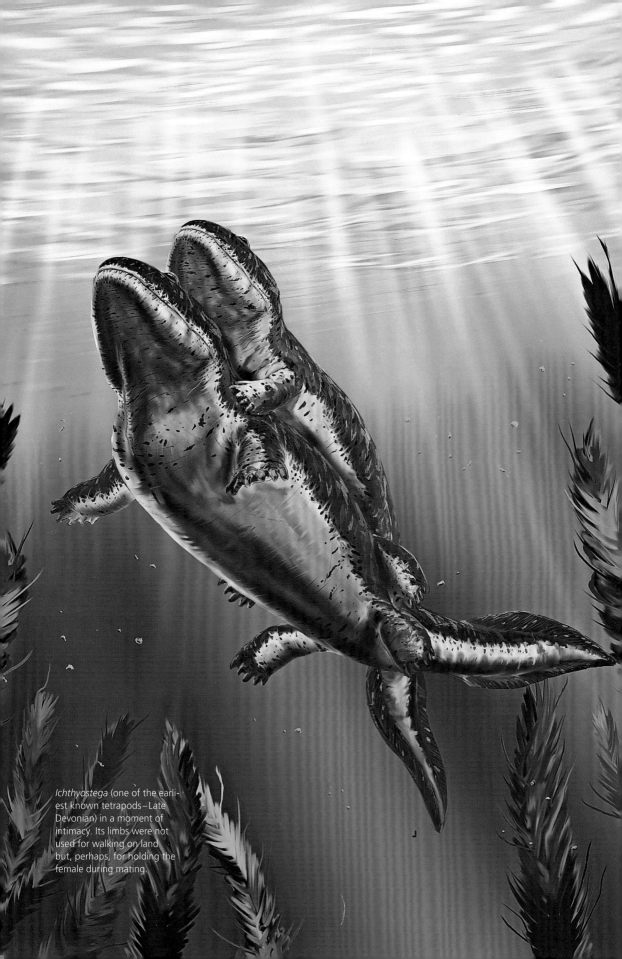

Ichthyostega (one of the earliest known tetrapods—Late Devonian) in a moment of intimacy. Its limbs were not used for walking on land but, perhaps, for holding the female during mating.

LIMBS: HOW DO THEY WORK? 2

LET US LEAVE FOR A MOMENT THE DEVONIAN AND THE MURKY waters inhabited by the early tetrapods. We are in the 1960s in a laboratory of teratological biology (the study of malformations that can appear during development). A surprising chicken embryo is the focus of our attention. The consequences of its mutation are such that the embryo is no longer viable, and will not reach full maturity. What does this mutant look like? It is so strange that fans of the comic *X-Men*, whose heroes have recently hit the big screen, might call it an "X-Chick": instead of wings and legs it is equipped with limbs that are much shorter than usual and have an excess of digits (fig. 2.1). Are you getting déjà vu? *Acanthostega* also had short limbs and was polydactylous (it had eight digits). And what is more, other mutant embryos even display beaks with tiny teeth. A mutant chick armed with teeth! This is again, without doubt, the re-emergence of an ancestral character, as the ancestors of birds (the feathered theropod dinosaurs) had teeth. If we take a closer look at the development of polydactylous mutant chicken embryos—a condition known as "talpid" because of their mole-like limbs (from "talpa"—Latin for mole), can we travel back in time through the course of evolution? Can we reconstitute the embryonic development that, 370 million years ago, culminated in the hatching of *Acanthostega* larvae? And can we sketch out, as a result, the sequence of developments that accompanied the evolution of chiridian limbs from the seven- or eight-digit condition of our Devonian ancestors to today's tetrapod limb, which never has more than five? Although there will always be uncertainties, there are clues which allow us to build such a scenario; if we can pinpoint the mutation(s) linked to the talpid pattern, we can even base that scenario on solid genetic foundations. This approach—drawing from observations of the development of living organisms to explain evolutionary and paleontological data—is relatively recent and is a discipline in its own right: "evo-devo" (short for "evolution-development").

In this chapter, we apply the evo-devo approach to the question of the emergence of chiridian limbs in Devonian tetrapods. Recall that if we observe the fins of ichthyan sarcopterygians close to the tetrapods (*Eusthenopteron*, *Panderichthys*, and their cousins) and the limbs of *Acanthostega*, in both cases we find first, starting from the bone attached to the girdle (scapular or pelvic), the humerus or femur; then the radius and ulna or tibia and fibula, followed by the wrist or ankle bones. However, whereas a fin terminates in dermal rays (the lepidotrichs), the chiridian limb extends into a hand equipped with digits. It is, therefore, the digits that define the chiridian limb.

> Let's ask ourselves what is known of the biological facts . . . why do our philosophers, thinkers and theologians claim, through pure speculation and divine inspiration, to have arrived at an understanding of the human organism? . . . Ignorance and superstition—these are the foundations on which men build their understanding of their own organism and its interaction with the external world; as for understanding embryological facts, this is entirely swept under the carpet.
>
> **Ernst Haekel (1834–1919)**

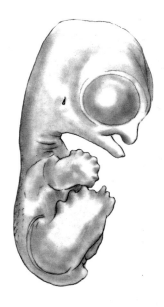

2.1. This "X-Chick" is a mutant which, instead of wings, has polydactylous limbs (with more than five digits!) like *Ichthyostega* and *Acanthostega* 360 million years before.

2.2. A mouse embryo in the ninth day of development: limbs start to make an appearance. At this stage, they consist of simple buds. The anterior bud (blue arrow) develops earlier than the posterior bud (yellow arrow).

Photo (scanning electron microscope) reproduced with the kind permission of Kathleen K. Sulik (Department of Cell and Developmental Biology, University of North Carolina).

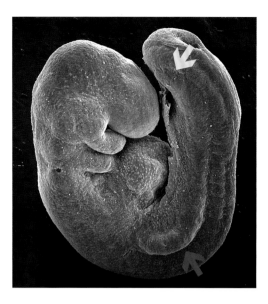

2.3. (facing) A mouse embryo in the ninth day of development, and a cross-section showing the organization of the anterior appendicular bud. This bud is formed following an intense proliferation of stem cells from a connective tissue, the mesenchyme. On the surface of the bud, the external dermal tissue (ectoderm) thickens.

Photos reproduced with the kind permission of Kathleen K. Sulik (Department of Cell and Developmental Biology, University of North Carolina).

Thanks to twenty-first-century technology, we can compare the sequences of events that bring about a chiridian limb and fin, bearing one question foremost in mind: How do digits form? Maybe this will give us hints to another question, asked in the first chapter: How did the tetrapods come into being?

Digital Birth

The origin of digits was, for many years, steeped in controversy. Toward the end of the nineteenth century, based on comparisons between sarcopterygian (for example, coelacanth) fins, and early tetrapod limbs, some paleontologists came up with the hypothesis that the two types of appendages derived directly one from the other. Digits were, therefore, interpreted as morphological transformations of the lepidotrichs. However, in the 1950s and 1960s, a Swedish paleontologist we are already acquainted with, Erik Jarvik, proposed a new hypothesis: digits are not derived from dermal rays of the fin but correspond to new evolutionary structures called "neo-formations." For thirty years the two interpretations faced off in the scientific arena – until the application of evo-devo allowed us to settle the matter.

Let us first look at the morphogenesis of a limb. The textbook case is that of the domestic mouse, *Mus musculus domesticus*, which is easy to breed and observe in the laboratory. In a mouse embryo, the morphogenesis of limbs follows a sequence of several stages.

Nascent limbs appear on the ninth day of embryonic development in the form of appendicular buds (fig. 2.2). Morphologically very simple and rounded, they come from an intense proliferation of stem cells from the mesenchyme (a spongy connective tissue derived from one of the three primary germ cell layers; the mesoderm). The mitotic activity is, therefore, intense within the bud (fig. 2.3). Distally, the ectoderm (external dermal tissue of the embryo) thickens and forms a slight bump on its surface.

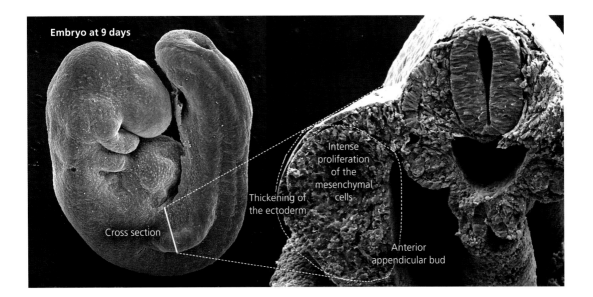

The second stage in limb morphogenesis is achieved on day 11 of the embryo's existence, when the appendicular buds are mitten-shaped (fig. 2.4). The ectoderm continues to thicken, especially along the distal rim of the bud, where it forms a small infolding; the apical ectodermal ridge (AER). In section, the cell infolding is located just above a small vascular canal.

It is the underlying mesenchyme that induces AER formation and which, in turn, via the secretion of small soluble molecules, induces the proliferation of mesenchymal cells. The AER is absent in limbless embryos – those tetrapods, including snakes, that lost their limbs over the course of evolution. In fact, its removal from the appendicular buds of a chicken embryo prevents the formation of legs or wings: the bud continues to enlarge but it does not lengthen, and limbs do not form; if removed

2.4. (below) A mouse embryo at day 11 of its development: the nascent limbs resemble mittens. At the distal tip of the bud, a strong thickening of the ectoderm forms an infolding: this is the apical ectodermal ridge, a decisive element in limb development.

Photos reproduced with the kind permission of Kathleen K. Sulik (Department of Cell and Developmental Biology, University of North Carolina).

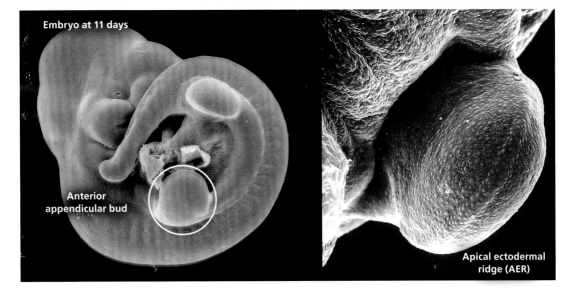

Limbs 45

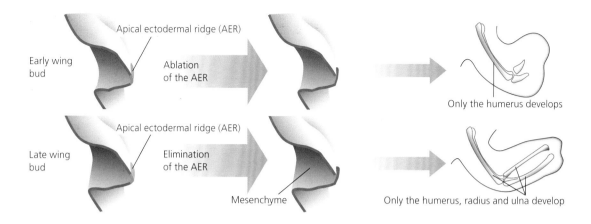

2.5. The role of the apical ectodermal ridge (AER), studied through experimental ablation. The AER is eliminated from the appendicular anterior bud of a chicken embryo at different stages of development. If the ablation is carried out early, only the humerus develops; if performed later, the radius and ulna form but not the digits. The AER is therefore indispensable to the lengthening of limbs during embryonic development. It stimulates cell proliferation in the underlying mesenchyme using small soluble molecules (growth factors).

early, only the humerus develops; if removed later, the radius and ulna form but not the digits (fig. 2.5). But when a second AER is grafted onto a normal appendicular bud, we observe the development of a supernumerary limb, more or less accomplished depending on the precision of the graft. The AER is therefore necessary and sufficient in the growth of the appendicular bud on the proximodistal axis—in other words, lengthwise growth—from the humerus to the last phalanxes.

On day 11 of embryonic development, the appendicular buds form a finely organized structure. The area located just below the vascular canal of the AER constitutes what is known as the "progress zone": this is characterized by a constant flux of mesenchymal cells which are "attracted" by those of the AER and actively proliferate due to the inductive effect of the latter. These cell movements take place from within the bud toward the surface, that is to say, in the proximodistal gradient. Therefore, the phalanxes are morphogenetically younger than the forearm bones, which, in turn, are younger than those of the arm.

Just below the progress zone, another remarkable cell region forms: this is the zone of polarizing activity (ZPA) which, experiments have revealed, orients limb growth along the anteroposterior axis—in other words, from the thumb toward the little finger (fig. 2.6).

At the same time, nerves and precartilage start to develop. The nervous system of the limbs is rooted in the central axis of the body (the thoracic region for the forelimbs, and the sacral region for the hindlimbs) and then radiates, somewhat like branches of a tree. The skeleton forms from islets of precartilage (hyaline spots) that appear without any apparent interconnection. These islets condense in the cell mass of the mesenchyme and will later give rise to endoskeletal elements of the stylopod (humerus or femur), zeugopod (radius + ulna or tibia + fibula) and autopod (carpals–metacarpals–phalanxes or tarsals–metatarsals–phalanxes). The number of islets is proportional to the available space—the more space, the more islets. Because the bud is smaller in its proximal part than in its distal part, it is in the latter that we find the majority of islets. Consequently, there will be more bony elements at the extremity of the limb (in the autopod) than at the base (the stylopod).

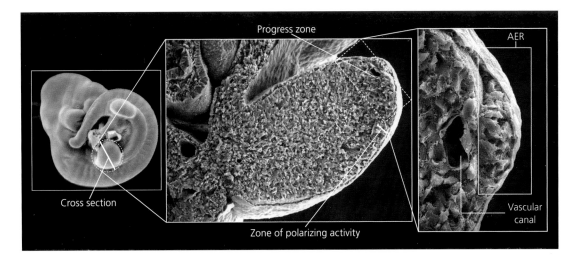

In the laboratory it is possible to vary the number of mesenchymal cells in the distal part of the chick's appendicular bud. If there are half as many, the space liberated allows the development of more precartilaginous islets and therefore more digits! A low count of mesenchymal cells seems, therefore, to induce polydactyly. The upshot is that the fewer mesenchymal cells there are in the autopod, the more digits there will be, and vice versa. The consequences of a diminished number of mesenchymal cells do not stop there: the cellular interactions in the appendicular bud are reduced (more "laborious"), the cells migrate at a slower rate toward the surface, and the limb takes longer to "grow" – it becomes stunted. Polydactyly and short limbs are the hallmarks of our "talpid" mutant but also, of course, of *Acanthostega*!

From day 12 the appendicular bud adopts a polygonal form, and looks much like a ping-pong paddle (fig. 2.7a). The increasingly angular outline can be explained by the presence of peripheral indentations: these are the digits that start to individualize following the disappearance of large cell regions between the precartilaginous islets. At this stage, cell destruction is an active process (known as apoptosis) while, at the same time, the extremities of the nascent digits are extending. The precartilaginous islets of the hands and feet develop by branching or by successive segmentation in such a way that a near-definitive skeleton of the hands and feet takes shape (fig. 2.7b and 2.7c).

We have seen so far that digits are formed because certain zones "grow" while others die. What happens, then, if we experimentally inhibit the process of cell destruction by apoptosis in the distal part of a limb in formation? There are two possibilities. If this inhibition is effective from day 11 of development, the precartilaginous islets will barely separate and we will observe polydactyly. If we intervene later in the course of embryo development, five digits form but remain attached to each other by webbing. This is exactly the condition we see in numerous web-footed tetrapods: ducks, for example, but also in *Acanthostega*. The equivalent for aficionados of science fiction would be the Man from Atlantis.

2.6. Appendicular bud of a mouse at the "mitten" stage (day 11 of development) in cross section. Two crucial regions for healthy limb development are visible: the apical ectodermal ridge (AER), essential for growth along the proximodistal axis (from the humerus towards the digits), and the zone of polarizing activity (ZPA), that orientates growth along the anteroposterior axis (from the thumb towards the little finger). Above the ZPA is the "progress zone," constantly crowded with cells from the mesenchyme. They are "attracted" there by the AER cells and actively proliferate thanks to the inductive effect of the latter.

Photos reproduced with the kind permission of Kathleen K. Sulik (Department of Cell and Developmental Biology, University of North Carolina).

Limbs   47

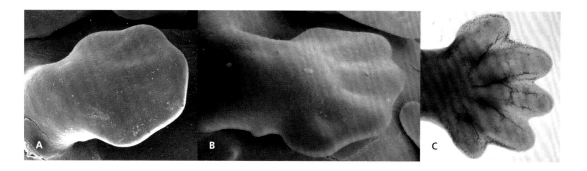

The last phase in the formation of a limb is the "web" stage (from around day 16 in mice, and later in humans). There are no longer any radical internal morphogenetic changes: the ossification of the autopods slowly achieves its end and the digits stop growing in such a way that only the proportions of the limb vary.

The Fold That Changes Everything

Now that we have seen the birth of a chiridian limb, let us confront the question of the emergence of a fin. Unfortunately the reference animal is rarely a sarcopterygian—there being very few laboratory dipnoi—but more often an actinopterygian (or ray-finned fish): the zebrafish, *Danio rerio*, is easy to breed but, at the same time, more distant from the "fish–tetrapod" transition.

In zebrafish, the beginning of the development of its paired fins resembles that of a mouse limb; we observe the formation of an appendicular bud and then a layer of ectodermal cells. Very quickly, however, the two morphogenetic scenarios diverge. Due to intense cell proliferation, the ectodermal layer develops a fold and finds itself separated from the rest of the bud by a gap (fig. 2.8) that is progressively colonized by cells, the provenance of which remains little understood, but from which the dermal rays of the fin originate. In the bud, once the fold is formed, the communication between the ectoderm and the mesenchyme is cut, and growth of the fin's skeletal elements (endoskeleton) ceases.

We can extrapolate that later formation of the ectodermal fold will prolong the proximodistal growth of the appendage, and the fin will consequently comprise more endoskeletal elements. Such differences in "timing" could thus explain why the pectoral fin of the zebrafish contains very few endoskeletal elements, whereas that of the coelacanth possesses bones homologous to those of the stylopod and zeugopod present in tetrapods. Furthermore, in coelacanths, the ectodermal fold does not form; the growth signals that continue to feed the apical ectodermal ridge allow cell proliferation that is at the origin of digit formation.

It is clear, therefore, that digits and lepidotrichs are distinct structures from an embryological standpoint. The conclusion we reach, through evo-devo reasoning, is that digits are indeed, from an evolutionary point of view, neoformations. They appear nearly 50 million years after the

2.7. The birth of digits. On day 12 of development in mice, the digits are just small peripheral indentations (A). They will then progressively individualize following the disappearance of large cell regions, giving way to inter-digital spaces (B). These are formed by a process of active cell destruction, apoptosis, as we can see in (C), where the cells marked with blue dye are undergoing apoptosis.

Photos reproduced with the kind permission of Kathleen K. Sulik (Department of Cell and Developmental Biology, University of North Carolina).

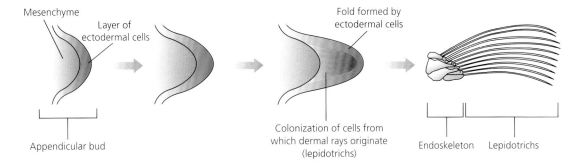

lepidotrichs without deriving from them (they are not fused or modified lepidotrichs). They are the result of a unique morphogenesis. This morphogenetic divergence is of fundamental importance because it expresses the main difference between fish (notably actinopterygians and ichthyan sarcopterygians) and tetrapods.

Hox Stories

One could be forgiven for thinking the transition from lepidotrichs to digits, the jump from fish to tetrapod, an easy stunt: as we have seen, it is just an apparently insignificant ectodermal fold that makes the difference. Nevertheless this jump has never yet been realized in a laboratory—neither genetic manipulation nor graft has managed to create a zebrafish (or any other fish, for that matter) equipped with digits. However, the genetics of digitation is far from being terra incognita and is even a particularly well studied scientific field, especially by the team of Swiss biologist Denis Duboule.

Among the genes that play a key role in the morphogenesis of chiridian limbs, homeobox (or homeotic genes) deserve special mention. These genes were discovered in drosophila (fruit fly) thanks to strange mutations that exhibited considerable morphogenetic "anomalies," such as the presence of well-formed organs located in another region of the body (the mutant drosophila *Antennapedia* possesses wonderful limbs instead of, and in place of, antennae!). These so-called homeotic transformations are the result of an "abnormal" (at a different moment or place) expression of homeotic genes during embryo development.

Our "model" tetrapod, the mouse, possesses homologues to the drosophila homeogenes. They are grouped in four complexes—HoxA, HoxB, HoxC, and HoxD—each one comprising between 9 and 11 genes (fig. 2.9). During the morphogenesis of a limb, it is the genes in complexes A and D—*Hoxa9, a10, a11, a13, d9, d10, d11, d12,* and *d13*—that play a primary role.

The deactivation of one of these genes results in the reduction or the loss of one or more skeletal elements in the limb (fig. 2.10). Thus, when the genes *Hoxa13* and *Hoxd13* do not function, the bones of the arm and forearm form normally, whereas those of the hand and digits are absent. If *Hoxa11* and *Hoxd11* are inactive, the arm is "normal"—the hands suffer only slight malformations—whereas the forearm itself is

2.8. Fin development in zebrafish. After the formation of an appendicular bud, cells in the ectodermal layer proliferate intensely—so much that they form a fold. This cuts communication between the ectoderm and the mesenchyme, preventing continued growth of the fin's endoskeleton. Digits cannot, therefore, form. On the contrary, the gap that separates the ectodermal fold from the mesenchyme is quickly invaded by cells from which the dermal rays (lepidotrichs) originate.

From Duboule and Sordino (1997).

Limbs 49

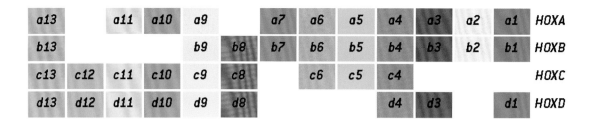

2.9. Hox genes (or homeotic genes): essential actors in the construction of an embryo. Each one expresses precise moments in embryonic development and in well-defined regions of the embryo. A mouse has four complexes (HoxA, HoxB, HoxC, and HoxD), each one containing about 10 genes, the result of duplication of a unique ancestral complex. The genes *Hoxa9, a10, a11, a13, d9, d10, d11, d12,* and *d13* are particularly involved in the morphogenesis of limbs.

2.10. (facing) The deactivation of certain homeogenes modifies the process of limb construction in mice. When the genes *Hoxa13* and *Hoxd13* do not function, hand and digit bones (autopods) are absent. On the other hand, if *Hoxa11* and *Hoxd11* are inactive then the radius and ulna are atrophied. If the four homeogenes *Hoxa11, Hoxd11, Hoxa13,* and *Hoxd13* do not function, then the limb is reduced to a stump: only the humerus forms. At the risk of oversimplifying, skeletal modifications induced by the inactivity of Hox genes impact along the proximodistal axis of the limb and in relation to the field and chronology of these genes' expression. This is, in turn, linked to their positioning on the chromosome (see fig. 2.9): the group 9 genes are expressed earlier and in areas more anterior and proximal than genes from group 13, expressed later and in regions more posterior and distal (after Zakany and Duboule [2007]).

atrophied. Finally, if we deactivate genes *Hoxa11, Hoxd11, Hoxa13,* and *Hoxd13* in our poor mouse, the limb is reduced to a stump: only the arm bones form.

It initially appears that skeletal modifications induced by the deactivation of these Hox genes are confined to the proximodistal axis of the limb. This observation is to be linked with the field and chronology of expression of the affected Hox genes which, in turn, depends on their distribution in the chromosome: the group 9 genes are expressed earlier and in anterior and proximal fields whereas group 13 genes are expressed later and in posterior and distal fields. Consequently, a gene such as *Hoxd9* will play a determining role in the construction of the arm, but *Hoxd13* will be the main player in the formation of the hand. There is another consequence of this template of Hox gene expression: the modeling of the hand and digits involves recruiting greater numbers of Hox genes (from groups 11, 12, and 13).

The reality, however, is much more complex; if we deactivate *Hoxd12* and *Hoxd13*, we observe an overexpression of *Hoxd11* in the autopod, resulting in polydactyly. *Hoxd11*, therefore, stimulates the formation of digits, a function normally counterbalanced by an inhibiting action exercised by *Hoxd12* and *Hoxd13*. We will go no further in our sketch of gene expression during the formation of a chiridian limb. Let us just retain the idea that the building of a chiridian limb is accompanied by a complex "musical score" in which different instruments—Hox genes—play at specific times and places. Armed with this knowledge, we turn our attention to the state of affairs in organisms lacking chiridian limbs but possessing fins, such as our old friend the zebrafish.

The four complexes—HoxA, HoxB, HoxC, and HoxD—are present, as are the same genes (even if their sequence is not exactly identical). There is, therefore, no "digit gene"—no more, for that matter, than there is a gene for any other organ or structure. A priori, *Danio rerio* has all the genetic potential to make digits. It is only the spatial and temporal dynamics of its gene expression that prevents this—for, while it employs the same "musical instruments" as a tetrapod, it is, in a manner of speaking, dancing to a different tune (fig. 2.11).

There is another element we need to bear in mind. The aforementioned Hox genes do not intervene uniquely in the morphogenesis of digits. They are also key actors in the formation of the urogenital system, the digestive tract, and the construction of certain skull bones. This raises several questions: Did digits appear in conjunction with the

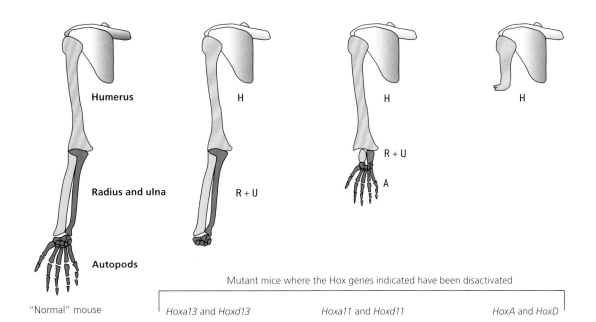

Mutant mice where the Hox genes indicated have been disactivated

"Normal" mouse | Hoxa13 and Hoxd13 | Hoxa11 and Hoxd11 | HoxA and HoxD

modification of one or more other body parts? Is there any link between cranial transformations involved in the fish–tetrapod "transition" (such as the modification of the branchial apparatus and transformation of the hyomandibular into the columella) and the appearance of chiridian limbs? So many questions and so few answers!

The complexity of these phenomena is, of course, a major obstacle, but there is also a problem linked to material. Let us not forget that geneticists work on living species that are often very different from those that inhabited the Devonian world. Researchers studying *Polyodon*, an actinopterygian slightly more basal than *Danio rerio*, reached similar conclusions regarding the "lateness" in the HoxD gene expression during fin development. The fact remains, however, that there is a world of difference between any actinopterygian you choose to mention and a sarcopterygian such as *Panderichthys* or *Tiktaalik*. The same goes for tetrapods; results obtained from a mouse or a frog are not necessarily applicable to *Ichthyostega* or *Acanthostega*. We are understandably a long way from knowing all the mechanisms involved in tetrapod origins.

Variations on a Limb

Until now we have talked of the chiridian limb. However, there is a striking difference between a mouse's paw and the front limb of a human, horse, chicken, or penguin. Nevertheless, in terms of organization – revolving around three elements: stylopod, zeugopod, and autopod – the development of different chiridian limbs follows paths almost identical to those described in a mouse: first buds, followed by mittens, then ping-pong paddles, and then webs!

In fact, the formation of each chiridian limb seems to be the result of variations on a theme: specialized zones (signal centers) control the formation of specific structures along the proximodistal axis (the apical

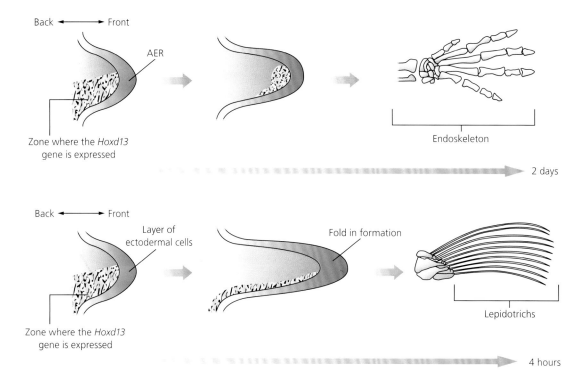

2.11. The different expression profiles of the same Hox gene (*Hoxd13*), comparing the embryogenesis of a limb (mouse) and a fin (zebrafish). Zebrafish and mice possess the same type of Hox genes. It is the dynamic of their expression in time and space that differs. There is, therefore, no "digit gene"! (after Duboule and Sordino [1997]).

ectodermal ridge), the anteroposterior axis (the zone of polarizing activity) and the dorsoventral axis (fig. 2.12); bones form from precartilaginous islets that are more numerous when more space is available in the mesenchyme; digits form from the precartilaginous islets and individualize by apoptosis (cell death at the origin of interdigital spaces).

So what determines the difference between chiridian limbs that are dissimilar both in relative part size and in number of digits? First, it is a matter of timing. From one species to another, the morphogenetic rhythm from different signal centers changes, engendering variations in limb proportions in every dimension (proximodistal, anteroposterior, and dorsoventral). This is a further reminder that morphogenesis occurs not in three, but four dimensions: it is like a music that, by altering its rhythm, alters its style. Second, the balance between the number of precartilaginous islets and cell death by apoptosis in the distal part of the limb in formation determines the number of digits. Consequently, there are many possible variations on a chiridian limb.

Let us look briefly at polydactyly. What is the maximum number of digits a hand or foot can bear? It is difficult to imagine having fifteen digits on each hand (and it would be inconvenient—except, perhaps, when playing the piano). In the living world a few tetrapods, such as frogs and the panda, have more than five digits; however, these are often not true digits (elements issuing from new structures) but the expansion of pre-existing phalanxes. We have seen in the "X-Chick" in the laboratory that it is possible to create polydactylous mutants, but, often, they do not

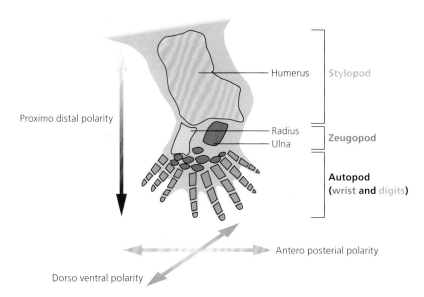

2.12. The forelimb of *Acanthostega*, with its three growth polarities. This drawing is an extrapolation based on findings from living tetrapods. From Devonian tetrapods to those living today—whether it be penguin, chicken, human, or any other—the formation of a chiridian limb involves specialized zones, called signaling cell centers, that control growth polarity of the limb along three axes: proximodistal, anteroposterial, and dorsoventral.

have true supernumerary digits. To find truly polydactylous tetrapods we have to turn to the past–first, to the Devonian. The "record," for the moment, is held by *Acanthostega* and its eight-digit hand. Its cousins are also polydactylous, but in different ways; in *Ichthyostega*, the supernumerary digits are anterior to digit 3 (preaxial polydactyly), whereas in *Tulerpeton*, they are posterior (postaxial polydactyly). A further flourish comes with the polydactyly of *Acanthostega*'s (eight-digit) hands, which are preaxial, whereas that of its (seven-digit) feet seems bilateral; that is to say that the supernumerary digits are "symmetrical" (occurring on either side of "normal" digits)!

Outside of the Devonian, are there other polydactylous tetrapods? The answer is yes; one of the ichthyosaurs (marine reptiles from the Jurassic of Europe and the Americas, known as ophthalmosaur) displays six true digits on each hand. This polydactyly appears to be bilateral and the hands of these surprising tetrapods also display polyphalangism (an increased number of phalanxes). In fact, the forelimbs of ophthalmosaurs are transformed into fantastic paddles (fig. 2.13). Another truly polydactylous marine amniote was discovered by Chinese paleontologists in the Triassic of China. It is equipped with seven digits in its forelimbs and six backing the hindlimbs. Its polydactyly is of a preaxial type, but apparently it does not display supernumerary phalanxes (fig. 2.14).

It is interesting to note that polydactyly appeared several times during the history of the tetrapods–in the Devonian, Triassic, and Jurassic–and in various forms. There seems to be only one constant: all the polydactylous fossils are marine, or at least euryhaline. Today, variations on the theme of digits are just as numerous without, however, exceeding five in number (barring "accidents"). Perhaps, in the course of evolution, certain combinations have proven more advantageous than others. In the living world, some digits are fused together (for example, the third and fourth

2.13. A surprising ichthyosaur: *Ophthalmosaurus* (Late Jurassic – 160 Ma). The name of this marine reptile comes from its large orbits. It also displays forelimbs transformed into six-digit paddles! In addition to this polydactyly, its hands are equipped with many supernumerary phalanxes (see inset).

combine to form a chicken's wing), and others have disappeared (in modern horses, the third digit – the central one – barely remains).

First Function(s)?

We now return to the Devonian. At the stage in history we arrived at by the end of chapter 1, around 370 million years ago, digits make their first appearance at the extremity of paired appendages in some sarcopterygians. We do not know yet how these digits emerged, and perhaps we never will, but we know that they are apparently neoformations – that is to say, they are not homologous to the dermal rays (lepidotrichs) of fins. This evolutionary innovation marked the birth of the tetrapods and the beginning of the chiridian limb's long journey.

How did this adventure all begin? We have seen that the limbs of *Acanthostega* were rather rigid and that their articulations allowed for only a limited range of motion. The forelimbs of *Ichthyostega* were apparently incapable of flexion and extension, and its paddle-shaped hindlimbs were oriented towards the back. Paleoenvironmental reconstitutions suggest that Devonian tetrapods lived in a landscape that resembled the mangroves of today. In short, contrary to previously popular and enduring belief, the early tetrapods were aquatic even to the point of being marine. So what was the function of those first freshly modeled chiridian limbs? On the shoreline their function would seem more or less limited to dragging the body out of the water. Jennifer Clack has put forward the idea that *Ichthyostega* could, once onshore, laboriously crawl around on the ground, much like present-day seals do. When it comes to aquatic functions, hypotheses are more abundant.

The chiridian limbs of early tetrapods are generally longer and better articulated than fins. They could allow movements that are impossible with fins, even in the water. It is probable that the first limbs contributed to maintaining bodily balance in the water (by acting as ballast), especially as the digits could play an essential role; the simple movement of a digit in a strong current would be enough to modify the trajectory of the body. It is easy to imagine *Acanthostega* on the surface, camouflaged among floating dead trunks and waiting for prey, its four limbs and digits splayed below the waterline acting as stabilizers; very much like crocodiles and newts today (fig. 2.15).

Most Devonian tetrapods lived in mangrove-like littoral environments, turbid and rich in plant debris. In the mangroves of today the mudskipper, a teleostean fish, is well known for its excursions onto land, using its robust pectoral fins to walk over roots. Once back in the water, its paw-like appendages allow it to maintain and support its body on the bottom. Did the first chiridian limbs have the same function? Could early Devonian tetrapods grip onto sunken branches and anchor themselves to the bottom? Perhaps. Digits have articulations that are very different from those of lepidotrichs; each is capable of independent movement. Did greater maneuverability play a role in the penetration, progression, and anchorage of those early tetrapods in the dense, dark, and stagnant environment they evolved in?

Stabilizers, hooks—we can imagine many other functions for the first chiridian limbs. "But [speculative] thoughts are cheap," writes Stephen Jay Gould. "Any person of intelligence can devise his half dozen before breakfast" (1991: 454). Now it is my turn, and I would like to put forward a hypothesis that is very simple but not often cited. What if the first chiridian limbs were used by the male to hold onto the female during mating? After all, the early tetrapods were amphibians in the broadest sense. Many present-day lissamphibians couple by external fertilization in the water; the male grips the female to deposit his sperm as close to her as possible. Perhaps *Acanthostega* and its Devonian cousins reproduced in the same manner. Of course all these hypotheses are merely speculation, to be, perhaps, confirmed or disproved in the light of more advanced biomechanical studies or new discoveries (for example, we still do not know what *Ichthyostega*'s hand looked like!).

This flood of hypotheses should be tempered with some sober reflections as we draw this chapter to a close. We will start with the obvious: it may be difficult for us to understand in our modern consumerist society, in which everything needs a sense and function (even our lives, deeds, and words), that a structure does not necessarily appear with a designated function. The evolutionary innovation that is the chiridian limb emerged subject to the whims of evolutionary contingency. As we have touched on in our lightning tour of the Hox genes involved in limb morphogenesis, it

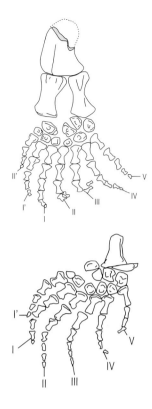

2.14. Hands and feet of a polydactylous (non-ichthyosaurian) marine reptile discovered in the Triassic of China. This amniote possessed seven digits in front and six at the back. Its polydactyly is of a preaxial type: its supernumerary digits are anterior to digit I.

Limbs 55

2.15. Swimming limbs and walking fins: (above) larva of a *Triturus marmoratus* using its forelimbs as ballast below the waterline; (facing) *Periophthalmus barbarus* using its pectoral fins to get around on land. Today's biodiversity abounds with examples that upset our anthropocentric vision of evolution ("fins for swimming" and "limbs for walking").

is perfectly possible to envisage digits as a consequence, a sort of secondary effect, of modifications in the urogenital system, the digestive tract, and the cranial skeleton in tetrapods.

We must also remember that an innovation can be linked to several functions. Evolution abounds with examples in which the principal function of an innovation in a living organism is not what it was originally intended for. Chiridian limbs appeared in the water, where we can imagine they played a whole host of functions (or possibly none at all), and only later proved "useful" for walking. It is a prime case of exaptation, meaning the deviated adaptation of an original function (if one existed), which is a common evolutionary process. In the dinosaurs, birds included, the feather probably did not appear for the purpose of flight. It quite possibly served initially as thermal insulation, allowing the animal to control its

body temperature (thermoregulate) and, secondarily – thanks to its highly branched structure – served to lighten the body. The feather would, therefore, be an exaptation for flight and an adaptation for thermoregulation. The same applies to the sarcopterygian lung, which could have initially been involved in the balancing of the body during swimming before it developed a true respiratory function. Put another way, the lung would be an adaptation to balance the body and an exaptation to breathe air. Evolution being what it is, learning the function of the first chiridian limbs may prove a fruitless quest.

THE PANGAEAN CHRONICLES 3

BY THIS STAGE WE ARE ALREADY ACQUAINTED with the ichthyan sarcopterygians, the closest relatives of the tetrapods, and we know what the early tetrapods looked like when they appeared in the Devonian (chapter 1). The major innovation that characterizes them is the chiridian limb. We have tried to understand, by comparisons between evolution and development (evo-devo), how this structure appeared in tetrapodomorphs (chapter 2). Contrary to a long-held idea, the early vertebrates equipped with limbs possessed more than five digits and were not terrestrial but aquatic. In this chapter we pursue our incursion into the evolutionary voyage of the Devonian tetrapods. We will discover how these species, at different times and with different approaches, installed themselves on terra firma. It is a complex story that unravels in both salt- and freshwater and it is a hard one to relate, it being so difficult to reconstitute the links between ancient tetrapods and those of today.

Devonian tetrapods are the very first representatives of a group dear to paleontologists: the stegocephalians (from the Greek "stego," plate, and "kephale," head). Under this label are grouped all the non-amniotic tetrapod fossils belonging to taxa no longer represented in the living world – in other words, all the fossil amphibians with the exception of the lissamphibians (anurans, urodeles, and caecilians). The stegocephalians were tied to the water through reproduction – they had to lay eggs there, and fertilization was probably external. Nevertheless, they comprised several species with a terrestrial lifestyle and it is from stegocephalian relatives that the first amniotes, organisms that were independent of the aquatic environment (see chapter 4), emerged during the Carboniferous Period.

The stegocephalians form a large group often characterized by flat skulls composed of a mosaic of flat, bony, dermal plates from which their name derives; by limbs furnished with a maximum of five digits (pentadactyl); and by the olecranon process, a bony extension of the proximal head of the ulna (cubitus), which is often found in tetrapods more derived than *Acanthostega*, and acted as a lever in limb mobility.

Even though they possess common characters, the stegocephalians do not form a clade: they do not share a unique common ancestor. The reference cladogram (see inside cover flap) clearly shows, for example, that their last common ancestor was also that of the amniotes. The stegocephalians constitute what experts call a grade. The term stegocephalian

Oh, the Polliwog is woggling
In his pleasant native bog
With his beady eyes a-goggling
Through the underwater fog
And his busy tail a-joggling
And his eager head agog –
Just a happy little frogling
Who is bound to be a Frog!

Arthur Guiterman

Plate Head

In the heart of Pangaea, 250 million years ago. Reconstruction based on analysis of the Moradi region of Niger.

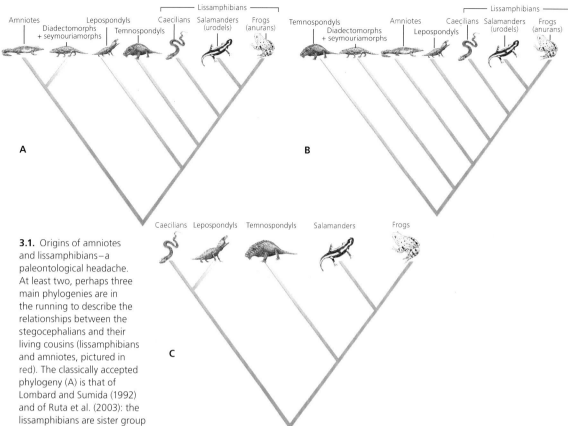

3.1. Origins of amniotes and lissamphibians—a paleontological headache. At least two, perhaps three main phylogenies are in the running to describe the relationships between the stegocephalians and their living cousins (lissamphibians and amniotes, pictured in red). The classically accepted phylogeny (A) is that of Lombard and Sumida (1992) and of Ruta et al. (2003): the lissamphibians are sister group of the temnospondyls, and the amniotes are sister group of diadectomorphs and seymouriamorphs. The comparative analysis (that is to say, the matrix taxon × characters) of Ruta et al. remains the most exhaustive to date. The phylogeny of Laurin and Reisz (1997) (B) suggests that the sister group of the lissamphibians is to be found amongst the lepospondyls. This hypothesis is fiercely debated because lepospondyls are highly specialized. Finally, the phylogeny of Carroll (2007) and Anderson (2008) (C), the most controversial, suggests that the lissamphibians do not form a clade; the caecilians would be sister group of the lepospondyls and the salamanders and frogs would be sister group of the temnospondyls. This contradicts all the other phylogenies, including molecular ones. Note that neither the lepospondyls nor the temnospondyls constitute a clade.

is conserved for practical reasons and is, at least, preferable to that of "labyrinthodonts" (a reference to the folded or labyrinthine structure of the dentine, a character also shared with ichthyan sarcopterygians).

The stegocephalian grade gathers several thousand species between the Late Devonian (the early tetrapods) and the Middle Cretaceous. Their fossils have been discovered on every continent, from the Canadian Arctic to Antarctica and from North America to Australia. It is remarkable to think that the stegocephalians, away from the limelight of media attention, reigned for some 270 million years—almost twice as long as the dinosaurs!

Between radiations and extinctions, stegocephalian evolution is like the musical score of a long symphony with most of the pages missing; paleontologists have a hard time making sense of it. It is worth the effort, as they may be considered the most enigmatic and difficult group to pin down in vertebrate evolution.

I will explain: the reference cladogram clearly shows that some stegocephalians are the closest taxa to lissamphibians (also called modern amphibians), whereas others are the closest relatives of amniotes. But which stegocephalians exactly? For the moment it is impossible to say with any certitude (hence the dotted lines on the cladogram). Whereas the origin of birds and mammals is increasingly well understood (thanks to fieldwork and advances in phylogenetic methodology in the laboratory), the question of amniote origins, especially that of lissamphibians, remains the number one black hole in vertebrate paleontology. As amniotes ourselves, the question touches the very heart of our existence: *Where do we come from?* For the moment there is no consensus as to the phylogeny of the stegocephalians within the scientific community.

A Controversial Phylogeny

Why is the phylogeny of the stegocephalians such a thorny problem for paleontologists? To answer this question, let us take a closer look at the diversity of the group. The reference cladogram shows us that it is made up principally of Devonian tetrapods (notably the *Ichthyostega* and *Acanthostega* of chapter 1), some strange Carboniferous forms (*Crassigyrinus* and *Spathicephalus*, which we will discover later on), as well as members of two groups: the batrachomorphs and the reptiliomorphs.

The reptiliomorphs include the stegocephalians more closely related to amniotes than are other stegocephalians, as well as the amniotes themselves. We will deal with them in chapter 4, when we get acquainted with the first amniotes (which were also the first reptiles) and the synapsids (the "mammal-like reptiles" and the mammals). Despite agreement among a growing number of paleontologists that amniote origins are to be found within a clade of reptiliomorphs called diadectomorphs, there are still sticking points on the stegocephalian tree.

Batrachomorphs comprise the stegocephalians more closely related to lissamphibians than are other stegocephalians, as well as the lissamphibians themselves. It is here that things get tricky–the reason being that frogs, salamanders, and other modern amphibians have a very specialized morphology. The skeleton of the frog, for example, is unique amongst vertebrates. Lissamphibians hardly resemble Paleozoic stegocephalians. They have their own morphological innovations. Although the latter sometimes help us to link lissamphibians together (intrarelationships), they do not provide information that would link them with other tetrapods (interrelationships).

The origin of lissamphibians therefore remains a mystery–a second sticking point on the stegocephalian tree. Discussions rage over which is the batrachomorph group closest to modern amphibians, temnospondyls or lepospondyls? The systematicians do not agree even on lissamphibian intrarelationships–that is to say, the relationship among frogs, salamanders, and caecilians. Figure 3.1 sums up the various points of view of the specialists. The phylogenies differ greatly depending on the author, and conference debates can get very animated!

Why So Much Ill Will?

Why is so much ink spilled on the origins of amniotes and lissamphibians? There are three main explanations for the divergent phylogenies.

The first is a question of methodology: all the authors use the same method, computerized cladistic phylogenetic analysis, but they do not input the same raw data (tables of morphological characters classified by taxon, or "taxon characters" matrix in cladist jargon); neither the morphological characters (choice, definition, coding, and polarization) nor the taxa (sampling and representativeness) are the same from author to author. It is hardly surprising that the end results should be different. Why this ruinous lack of agreement among phylogeneticists? Rampant competitiveness? A race for publication? It is a mixture of both. The prevailing ambience among researchers is, unfortunately, one of competition more than cooperation.

The second problem owes itself to the nature of the taxa to be classified: each of them, as we have seen, is highly specialized and their specific characters are often uninformative when it comes to studying relationships among them.

The last problem is a lack of data on development, notably larval, of the species being compared. We saw in chapter 2 how development (ontogeny) can sometimes provide precious information about evolution (phylogeny). This lack is glaring in fossil species, of course, as the preservation of immature individuals is rare. More surprisingly, this is also a problem with certain living tetrapods (caecilian amphibians, for example) that are difficult to raise or observe in their natural habitats.

There are, nevertheless, solutions to these problems. Paleontologists have at their disposal all the necessary tools to share data and merely require a simple internet forum in which to come to agreement on the definition of given characters, or a common online database in which to establish a sufficiently standardized matrix. There exist today a sufficient number of alternatives to share knowledge (or at least raw data) without systematically resorting to screening committees or the reviewers of scientific journals. As Stephen Jay Gould notes,

I know from my own experience as a participant in major scientific debates that the explicit record of publication is utterly hopeless as a source of insight about shifts, forays, and resolutions [which characterized the discussion]. As Peter Medawar and others have argued, scientific papers . . . are, at best, logical reconstructions after the fact, written under the conceit that fact and argument shape conclusions by their own inexorable demands of reason. Levels of interacting complexity, contradictory motives, thoughts that lie too deep for either tears or even self-recognition – all combine to shape this most complex style of human knowledge. (1987: 85)

Next, researchers must be given the means to carry out fieldwork. There are no fossils without fieldwork and, without fossils, no paleontologists. The avowed goal of paleontologists working on stegocephalians is to find new taxa in Paleozoic rocks to fill in the blanks. It would be interesting to discover Paleozoic lissamphibians less specialized than Mesozoic forms and, in an ideal world, well preserved and showing different growth stages (larval, juvenile, and adult).

Crisis Time

JUST AFTER THE APPEARANCE OF THE EARLY TETRAPODS (which were also the first stegocephalians), a massive life crisis affected largely marine environments at the limit between two Late Devonian stages, the Frasnian and Famennian (370 Ma). Together with the Permian–Triassic extinction event (250 Ma), the Frasnian–Famennian event is one of five great mass extinctions that have marked Earth's history ("the Big Five" in paleontology!). Is this crisis linked with the emergence of early tetrapods from the water? Paleontologists do not know yet.

If we jump back in time 365 million years, the face of the Earth is very different from that of today, and the atmosphere very different from the air we breathe today. The continents, mainly clustered together in the Southern Hemisphere, are surrounded by oceanic masses harboring strange armored fish, corals, and marine arthropods. On land, the first fern and horsetail forests are springing up. These are home to numerous terrestrial arthropods but, as yet, no terrestrial tetrapods (several million years will elapse before they leave the water). The climate changes brutally and, globally, high temperatures give way to an ice age.

This climatic shift has a devastating effect on the fauna: most benthic organisms are decimated, including corals, stromatoporoids, brachiopods, and trilobites. The continents also become a hecatomb: some plant groups die out, and the composition of others is radically altered. The mechanisms and exact causes of this catastrophe remain little understood. It is true to say that the Frasnian–Famennian extinction was rapid on a geological time scale. The climatic hypothesis is favored for the continental masses (warming followed by global cooling due to glaciation). In the seas the phenomenon of mass extinction appears linked to a period of general anoxia (a dramatic drop in oxygen levels in the water). Glaciation and anoxia are not necessarily mutually exclusive: forests ravaged by plunging temperatures can discharge enormous amounts of lignite debris and sediments on the coast and in the seas, perhaps saturating epicontinental waters and asphyxiating the animals that live there. Is this what happened? The question remains open; but it is not because two phenomena are contemporary that they are necessarily linked. Put another way, correlation is not a synonym of causality.

One thing is certain: an enormous glaciation traversed southern Pangaea from the Late Devonian to the Early Carboniferous—a veritable Paleozoic ice age!

No Such Thing as Leaving the Water!

Even though the phylogeny of the stegocephalians remains controversial, it does not prevent us from formulating hypotheses about their evolution and the transformations they underwent throughout their long history. As we saw in the first chapter, the early tetrapods (that is to say, the first stegocephalians) experienced a formidable evolutionary radiation from their beginnings in the Late Devonian (365–370 Ma). In several million years the group "exploded"—diversifying and thriving just about everywhere (their fossils can be found from North America to China). The apparition of the chiridian limb was a resounding success.

Devonian stegocephalians, including the very first ones such as *Ichthyostega* and *Acanthostega*, were aquatic. Most lived in brackish waters, and even marine environments. They made it through one of the five life crises that this planet has known (the Frasnian–Famennian extinction event; see inset "Crisis Time"). It is only later, during the Early Carboniferous (330–335 Ma), that certain forms left the water. Afterward, the music speeds up and the evolution of the stegocephalians takes on epic proportions on land as well as in the water.

Terrestrial or Aquatic—Everything's Relative

It needs to be stressed from the outset that the expression "leaving the water" is not obvious when applied to the stegocephalians—that is to say, non-amniotic tetrapods linked to the water, at least for reproduction. In fact, the lives of many of them are difficult to fathom. These vertebrates often evolved at the interface between land and water. By their morphology, we can easily distinguish typically terrestrial stegocephalians (which apparently only returned to the water to reproduce and lay eggs) from the typically aquatic (which spent most of their time in the water) and come up with hypotheses about their lives. The same does not apply to the amphibious forms that may have made up the majority: it is difficult to estimate how they split their time between land and water.

The notion of terrestriality therefore becomes relative. Some paleontologists think that as soon as a vertebrate (whether ichthyan or tetrapod) is capable of venturing onto dry land we can talk of leaving the water. In these terms, the mudskipper, a remarkable living fish that can climb mangrove roots, embodies an attempt to leave the water happening before our eyes. To others, leaving the water is only accomplished when a tetrapod becomes entirely independent of the aquatic environment—that is to say, with the invention of the amniotic egg and the appearance of the amniotes (see chapter 4).

Several Sorties from the Water

The expression "leaving the water" is false because it is in the singular: it is meaningless to speak of *the* leaving of the water to describe an evolutionary episode that saw the early tetrapods slide a foot out of the aquatic environment.

As we already know, in the history of life, organisms have never stopped leaving the water! Between the Ordovician and the Silurian (440 Ma), plants and then arthropods embarked on a terrestrial adventure, and with some success (see inset "Plant and Arthropod Excursions onto Dry Land"). Today, also, aquatic organisms venture onto the continents. A case in point is the coconut crab *Birgus latro*, which spends much of its time out of the water. Most principal living clades (for example, terrestrial gastropods, whose ancestors were probably aquatic) have terrestrial forms and therefore their own "leaving the water." Without multiple plant and arthropod departures from the water—which considerably modified Paleozoic landscapes, ecosystems, and environments—the tetrapods would probably have never ventured onto land.

Next, several sorties from the water took place within the tetrapod group itself (as among plants and arthropods). As we can see on the reference phylogenetic tree (see inside cover flap), several lineages ventured onto dry land. One of the first stegocephalians to leave the water (a somewhat timid effort; we must not forget that, for these amphibians, water was home, at least for reproduction), was probably *Balanerpeton* in the Early Carboniferous. A little later, during the Late Carboniferous–Early Permian (300 Ma), the stegocephalians discovered new continental ecological niches and experienced several evolutionary radiations. Chapter 4 treats those of reptiliomorphs and early amniotes. In this chapter we concentrate on the radiations affecting the temnospondyls (a group, of which *Balanerpeton* is one of the first representatives, that dominated the end of the Paleozoic) and the lepospondyls. From the Carboniferous to the Permian the lepospondyls evolved discretely, side by side with the temnospondyls. We will see that these two groups of batrachomorphs comprised of both terrestrial and aquatic forms, the latter likely the result of readaptation to the aquatic environment! We will also see that some lepospondyls—which were terrestrial, according to some paleontologists—had already lost their limbs.

It would seem that all these Late Carboniferous–Early Permian "exploits" were favored by continental environments and climates propitious for new life forms. At that time—and notably in equatorial latitudes—a large part of Pangaea was covered in deep forests, dense and luxuriant, a little like the Amazon jungle today, but without flowering plants. Landscapes of giant horsetails, conifers, and calamites—already host to a multitude of terrestrial invertebrates—were the stomping ground and food source for the first stegocephalians and early amniotes. These forests gave us today's great coal deposits, stretching from North America to the Czech Republic, passing through France and Germany. The reign of the temnospondyls was to last until the end of the Early Cretaceous. These stegocephalians would rub shoulders with the first lissamphibians, which emerged in the Early Triassic (250 Ma).

Plant and Arthropod Terrestrializations

IN THE PLANT WORLD, MOSSES (BRYOPHITES) ARE AMONGST the first terrestrial species. They possess a new structure, the sporangium, which contains spores, facilitating their dispersal into the air at maturity. The oldest known mosses date back 475 million years (only their spores have been fossilized). Before them, in the fossil record, other plants were capable of tolerating desiccation. The coleochaetes, for example, are strange encysted green plants able to conserve and nourish their "germs." The oldest complex, vegetative soils (paleosoils dating from the Late Ordovician, about 450 Ma) were apparently covered with hepatics (or liverworts—plants close to mosses; only their spores have been found), but older paleosoils—1 billion years old!—containing multicellular cysts and thalli have been recently described in Scotland. The lichens, marvelous examples of symbiosis between algae and fungi, started to settle on dry land a mere 400 million years ago, if not earlier. Older lichens (around 500 million years old) have been discovered in China but belonged to a shallow marine environment.

After lichens and mosses, other plant groups settled on land, each group developing its own innovations: roots that could tap the water and mineral resources of the soil, a waxy cuticle to protect from desiccation, lignin in the wood that maintains the plant vertically, or stomates on the leaves that participate in transpiration (water expulsion) and gas exchanges (photosynthesis and respiration). By the end of the Devonian, the vegetal groundcover was almost as developed as that of today. All the structural types of terrestrial plants—arborescent, bushy, liana, epiphytic, or any other—were already present. It is only the species and their reproductive systems that have changed between the Devonian and today, with the appearance of the large groups of seed-bearing plants (conifers and

3.2. A rocky coast 400 million years ago. At that time there were no land vertebrates. Nevertheless, there had already been several terrestrializations. Mosses were amongst the first plants to venture onto the continents more than 500 Ma (according to recent discoveries in Scotland). The first terrestrial lichens, shown here, appeared several million years later. At the end of the Devonian, 360 Ma, the structure of the vegetal ground cover was almost as developed as that of today.

Earth before the Dinosaurs

angiosperms). As a comparison, imagine if, by the end of the Carboniferous, all continental, terrestrial, and aerial environments were already occupied by the tetrapods.

The arthropods followed the wave; the vegetal land cover and subsequent formation of soils proved very favorable to these previously mainly aquatic animals. Like plants and tetrapods, they made several exits from the water during evolution: myriapods in the Silurian, chelicerates in the Devonian, and crustaceans and insects later. The first truly terrestrial arthropods were little myriapods from the Late Silurian (420 Ma). Detrivore tending towards fungivore (mushroom munchers), they lived in the soils. The terrestrialization of arthropods may go even further back: fossil traces attributed to terrestrial arthropods have, in fact, been identified in the Ordovician, before even the first plants (mosses and hepatics).

One thing is sure: by the Carboniferous period the dense, humid forests were seething with a very diverse entomofauna. Some forms were of a size that defies belief: pluricentimetric cockroaches, dragonflies with a 70-centimeter wingspan (*Meganeura*), giant spiders (*Megarachnides*), millipedes several meters long (*Arthropleura*)! These carnivorous and/or detrivorous arthropods had practically no predators. Absence of pressure from competition might explain these giant forms.

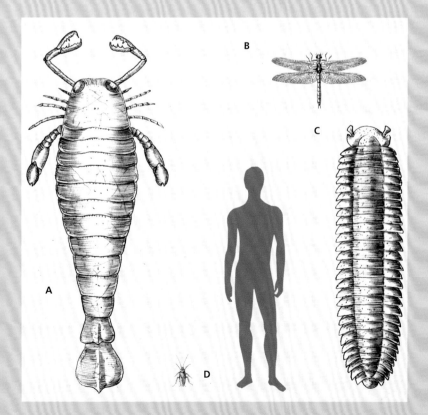

3.3. Giant arthropods of the Carboniferous coastal forests: (A) *Eurypterus* (scorpion); (B) *Meganeura* (dragonfly); (C) *Arthropleura* (millipede); (D) *Spiloblatta* (cockroach). The human silhouette is for scale. The near absence of predators and a high atmospheric oxygen level might have permitted these arthropods to attain record dimensions.

This is a story with many twists and turns, in which the exploration of new environments is linked to an explosion of new and amazing forms. From the Carboniferous Period onward, some stegocephalians present morphologies that can only be described as stunning.

The Enigma of the Carboniferous

We still have little information about Early Carboniferous stegocephalians; this is not the case for those of the Late Carboniferous. It seems that this scarcity of fossils is a taphonomic problem: the lack of Early Carboniferous continental outcrops limits the potential for discoveries (this is Romer's Gap). Moreover, the few tetrapods known from this time are very different from each other. This is a real headache for my colleagues who try to classify them.

Among the aquatic stegocephalians we shall meet strange forms such as *Crassigyrinus* or *Spathicephalus*. On the terrestrial front, several trackways discovered at Horton Bluff, a very well known Canadian fossil site (in New Scotland), show us that some Early Carboniferous stegocephalians were already attaining respectable sizes, with limbs up to 20 centimeters in diameter! Terrestrial stegocephalians from the same period have also been found in Europe, and vary greatly in size and shape. First in the spotlight is *Balanerpeton*.

Run for Your Lives! *Balanerpeton woodi*

Balanerpeton woodi was discovered at the well-known Scottish site of East Kirkton, which dates from the Early Carboniferous (335 to 330 Ma; fig. 3.4). About thirty pluricentimetric skeletons, in curled-up position, were found in the hollow interiors of fossil lycophyte tree trunks (a group to which present-day sigillarians belong). This suggests that these were the last refuges these small amphibians found to protect themselves from a mudslide that proved fatal (fig. 3.5).

The morphology of *Balanerpeton* is very peculiar: it possesses a large tympanic opening at the back of the head (the otic notch), a window for sound detection outside the water. Its wrists and ankles, well ossified and robust for the size of the animal, suggest a terrestrial locomotion. Furthermore, *Balanerpeton* possesses neither lateral line nor the bony gills that characterize an aquatic life. All the anatomical arguments converge to indicate that *Balanerpeton* spent most of its time on land. This fossil has earned itself more than one title: it represents not only one of the oldest temnospondyls, but also one of the first terrestrial tetrapods.

With its rounded skull and its fairly long body, *Balanerpeton* looked like a salamander. It also represents one of the rare fossil amphibians that possess almost twice as many teeth on the upper jaw (small and 40 in number) as on the lower jaw. This very special pattern, called "dignathic heterodonty," suggests a particular diet—carnivorous tending towards insectivorous.

Balanerpeton

Age Early Carboniferous–335 to 330 Ma
Location Scotland, United Kingdom
Size Up to 20 cm
Features Flat, rounded skull; well-ossified limbs
Classification Basal temnospondyl

3.4. *Balanerpeton woodi* (335–330 Ma): (A) skull in palatal view with part of the vertebral column; (B) reconstruction of the skeleton in dorsal view; and (C) skull in palatal view. This fossil, which owes its name to the famous collector Stan Wood, was described in 1994 by the British paleontologist Andrew Milner. Despite having to return to the water to reproduce and lay eggs, this Early Carboniferous stegocephalian is one of the very first terrestrial tetrapods. *Photo courtesy of Jeff Liston, Hunterian Museum and Art Gallery, University of Glasgow.*

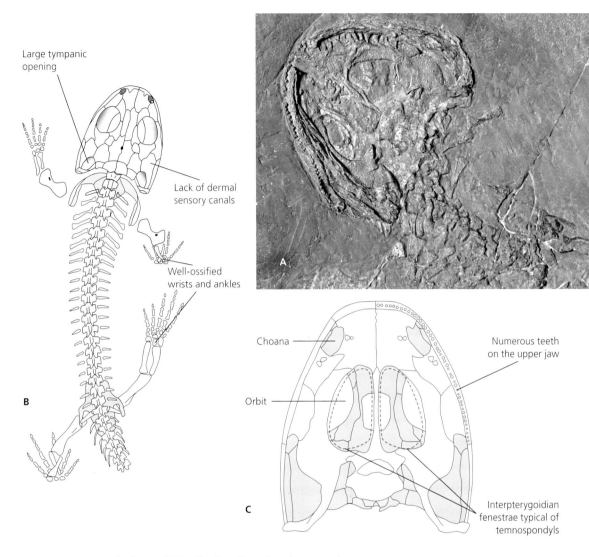

A Morphological Headache: Crassigyrinus scoticus

While some stegocephalians experimented with a terrestrial existence, others evolved in an aquatic environment. This was the case with the astonishing *Crassigyrinus scoticus* from the Early Carboniferous of Scotland. This fossil was found in layers aged around 325 million years (Viséan–Namurian), in ancient mines near Edinburgh. It is a pity that today the mines are closed; the anatomy of this sort of giant tadpole, two meters in length, has been puzzling paleontologists for more than 20 years. They would love to go there and look for more specimens! The evolutionary history of *Crassigyrinus* is also rather murky.

With its imposing skull and atrophied limbs (especially the front ones, in which the humerus is the same size as the orbits), *Crassigyrinus* would have been an excellent swimmer that, according to the type of sediments in which it was discovered, lived in freshwater (fig. 3.6). Its basin

Crassigyrinus

Age Early Carboniferous–325 Ma
Location Scotland, United Kingdom
Size Up to 2 m
Features Imposing skull; very reduced limbs; well-developed caudal fin
Classification Basal tetrapod

3.5. A devastating mudslide 330 Ma. It sweeps away individuals of one of the oldest terrestrial tetrapods: *Balanerpeton woodi*. We think this is what happened to the 30 skeletons that were found in a curled-up position inside fossil tree trunks.

and vertebral column were little ossified, probably facilitating movement by undulation, as marine snakes move today.

Crassigyrinus has an unusual skull composed of highly ornamented bones, a bit like its temnospondylian cousins (see below) or those of living crocodiles. The ornamentation pattern of these cranial bones is irregular, suggesting a supple skin or hide, perhaps sporting protuberances. With such attributes, *Crassigyrinus* probably excelled in the art of camouflage. This amphibian also presents a high, short snout. Carnivorous tending toward piscivorous, it was armed with multiple fangs inside the mouth and on the palate. More impressive still, its lower jaw had great mobility, a feature observed in living snakes that allows ingurgitation of prey sometimes bigger than themselves. The ugly *Crassigyrinus* lived in turbid water rich in plant debris, and it probably hunted by stealth, all but invisible in a labyrinth of roots.

Given all these very strange particularities and its highly specialized aquatic morphology, *Crassigyrinus* remains a true enigma. Previously considered to be close to the reptiliomorphs, notably based on its skull ornamentation, *Crassigyrinus* is now (and perhaps only until the next phylogenetic analysis) the most basal Carboniferous tetrapod: in the reference cladogram it appears just after the Devonian stegocephalians.

Frisbee Head: *Spathicephalus mirus*

A Frisbee is certainly the first thing that springs to mind when describing *Spathicephalus mirus*, a ludicrous-looking stegocephalian with a skull shaped like a flat disc. From the Early and Late Carboniferous of Canada and the United Kingdom, this amphibian of modest size (its skull has a diameter of around 20 centimeters) perfectly illustrates the different

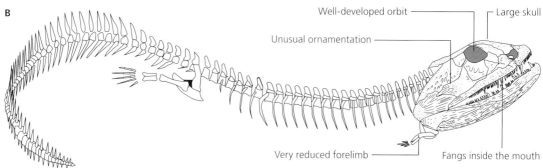

3.6. (A) *Crassigyrinus scoticus* (325 Ma) hunting by stealth, hiding amongst the roots in freshwater. (B) Reconstruction of the skeleton. With its numerous fangs and highly mobile lower jaw, this stegocephalian must have been a formidable predator.

3.7. (facing) (A) *Spathicephalus mirus* sifting through the mud. (B) Skull in dorsal and lateral right views. Known from the Early Carboniferous of New Scotland (Canada) and the Early and Late Carboniferous of the United Kingdom, *Spathicephalus* was a strange aquatic amphibian with a "Frisbee head." Its strange dentition suggests it was a filter feeder.

morphological experiments or "anatomical DIY" undertaken by the early tetrapods (fig. 3.7).

With a circular, flattened skull that almost appears two-dimensional, *Spathicephalus* would have been aquatic. This is also suggested by the sediments, typical of a freshwater environment, in which it was found. On the dorsal surface of the skull, there are strange bean-shaped orbits set very close to each other. Did *Spathicephalus*'s eyes take up all that space? Probably not.

These orbits perhaps also housed, as well as ocular globes, glands or electrosensory organs with which our amphibian could detect the movements of predator or prey in the vicinity. Such organs are present in living forms that spend much of their time lying buried in the mud, as perhaps was the case with *Spathicephalus*. Another hypothesis is that there were muscles inserted in these deep cavities, allowing this stegocephalian to close its jaws rapidly on its prey. Its very flat jaws would otherwise be slowed by increased water resistance.

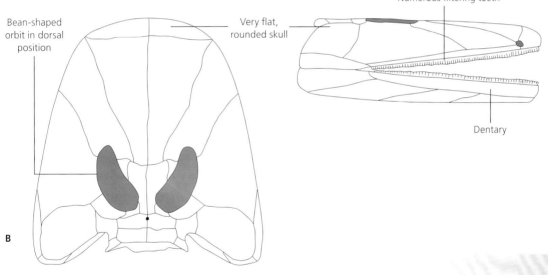

Bean-shaped orbit in dorsal position

Very flat, rounded skull

Numerous filtering teeth

Dentary

Another enigmatic characteristic of *Spathicephalus* is its strange dentition: it had numerous very delicate, tiny teeth, but no fangs. The dentary alone (lower jawbone) holds a record 120 teeth! The extremities of these teeth are pectinate, that is to say, comb shaped. This suggests that *Spathicephalus*, the "Frisbee stegocephalian," was a filter feeder, meaning that it fed by swallowing large quantities of water and mud that was then filtered with the aid of "dewlaps" to extract small invertebrates or fish. With its comb-like dentition, *Spathicephalus* is a case apart.

Spathicephalus

Age Early and Late Carboniferous
Location Canada and United Kingdom
Size Skull 20 cm in diameter
Features Round, flat skull; numerous small filtering teeth
Classification Basal tetrapod

The Comings and Goings of Temnospondyls

Eryops

Age Permian
Location United States
Size Up to 2m
Features Long snout; robust girdles; large, overlapping ribs
Classification Eryopoid temnospondyl

3.8. (A) The classic temnospondyl: *Eryops megacephalus* (Permian of North America) is pictured here in an encounter with *Dimetrodon*, a carnivorous synapsid ("mammal-like reptile") in the background. (B) The entire skeleton (lateral right view) and skull (palatal view). The temnospondyls formed an important group of stegocephalians, occupying aquatic and terrestrial ecosystems for nearly 250 million years – longer than the dinosaurs!

We have already mentioned *Balanerpeton*, which appeared in the Early Carboniferous (Viséan) with one of the most important fossil amphibian groups that the Earth has ever harbored – the temnospondyls. This group of carnivorous opportunists deserves a second glance.

Temnospondyls quickly occupied a wide variety of environments and took on terrestrial, amphibian, and aquatic forms. Latest phylogenies suggest that the aquatic forms shared a common ancestor that was rather terrestrial. This means that temnospondyls returned to the water during their evolution and the chiridian limb, which appeared in early aquatic Devonian tetrapods (notably *Balanerpeton*) was then "used" on land and later "recycled" in the water. Water, land, water: stegocephalian evolution is a constant back and forth between two phases.

From the end of the Carboniferous to the beginning of the Permian, the temnospondyls spread across the entire planet: their fossils are to be found from Antarctica to Norway and from North America to China. They colonized every environment linked to water: swamps, flood plains, streams, rivers, lakes, deltas, coasts, and even the open seas. By the end of the Paleozoic – before dinosaurs, birds, and mammals – the temnospondyls reigned supreme. Five hundred species strong – maybe more – they are the most frequently found vertebrates in many layers dating from this period. Diverse in appearance – some like large salamanders, others like ferocious crocodiles – they constitute a world in their own right. A world where certain species dominated others, with 8 meters to quash any dissension!

The evolutionary history of the temnospondyls is complex and punctuated by expansive radiations and extinctions. The group crossed several major life crises including the most murderous of all, marking the boundary between the Late Permian and the Early Triassic 250 million years ago (see chapter 4). This period is marked by a global faunal "turnover" within the temnospondyls: some forms die out while others appear. The group definitively disappeared from the fossil record in the Albian (late Early Cretaceous). They reigned for 220 million years – a tetrapod record (let us not forget that non-avian dinosaurs only lasted 135 million years)!

After presenting the morphological characteristics of the temnospondyls, we will pick out some choice moments in their evolution. First of all, let us find out about those which are, according to most paleontologists, close to modern amphibians (lissamphibians). We will then discuss the surprising aquatic readaptation that some underwent. We will then learn about the largest amphibian of all time (approaching 10 meters) and finish by painting a picture of the temnospondyls that, some 250 million years ago, flourished at the heart of Pangaea, in the present-day Sahara.

A Group of Character and Distinction

The temnospondyls (from the Greek "temnos," (in) several (parts), and "spondylus," the vertebra) are characterized by vertebrae in several parts:

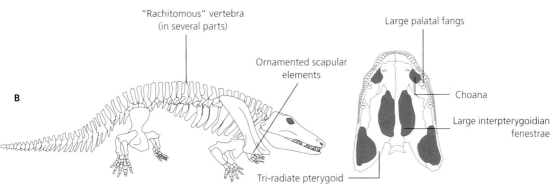

a dorsal part extending into a blade (the neurocentrum), and often two ventral parts around the spinal cord (the pleurocentrum and the intercentrum). Such vertebrae are also called rachitomous. In some Mesozoic forms, however, there are only two parts, one dorsal and one ventral, which form a disc.

3.9. (A) *Doleserpeton annectens* – 285 Ma. Is this temnospondyl from the Permian of North America the closest relative of all living amphibians (lissamphibians)? Paleontologists differ on this question. (B) In any case, *Doleserpeton annectens* is interesting because its small pedicelate and bicuspid teeth evoke those of lissamphibians and some lepospondyls. (C) Skull in dorsal view.

Another characteristic of temnospondyls is the skull ornamentation: the dermal bones that form the skull roof have a complex network of small bumps and alveoli resembling a honeycomb. It is true that the same pattern exists in other stegocephalians and living crocodiles, but this type of ornamentation on the scapular girdle is peculiar to the temnospondyls.

They also all have large fangs at the front and inside the palate ("palatal fangs") and large ovoid windows set further back (interpterygoidian fenestrae), which often represent half the skull size and are bordered by triradiate bones called pterygoids (fig. 3.8): this battery of characters would have made the temnospondyls redoubtable predators. It is easy to imagine a victim suddenly caught and immobilized in the jaws (the interpterygoidian fenestrae serving as a powerful buccal pump) and completely skewered on the palatal fangs.

At the Origin of Modern Amphibians? Doleserpeton annectens and Amphibamus grandiceps

The phylogenetic position of the temnospondyls places these stegocephalians at the heart of a burning issue—the origins of lissamphibians. According to a majority of paleontologists, the sister taxon of the lissamphibians should be somewhere among the dissorophoids, a very peculiar temnospondyl group principally characterized by highly fenestrated skulls (with large orbits, voluminous interpterygoidian fenestrae, and an impressive pineal foramen) and limb bones that, when preserved, are often slender and long. Two dissorophoids from the amphibamid family are often put forward as the closest relative of the lissamphibians: *Doleserpeton annectens* and *Amphibamus grandiceps*.

Doleserpeton annectens is an unusual, small stegocephalian from the Early Permian (285 million years) discovered in the Dolese quarry in Oklahoma (fig. 3.9). Representative of a rather terrestrial fauna, it bears pedicelate (with a reduced base or articulated on a peduncle) and bicuspid (two-crested) teeth, as do most lissamphibians. Does this character link *Doleserpeton* to the lissamphibians, or is it simply an evolutionary convergence due to a particular pattern of feeding or prehension of prey and which appeared several times in vertebrate history? We will see that lepospondyls also had this type of dentition.

It was American paleontologist John Bolt who was the first to notice and publish in 1969 on this striking similitude (although Cope, in 1888, associated the temnospondyl group with living amphibians). Today, according to Marcello Ruta's computerized phylogenetic analysis, one of the most exhaustive ever carried out (see fig. 3.1), this morphological resemblance is a clear sign of a close relationship with lissamphibians.

Amphibamus grandiceps (fig. 3.10) is also a dissorophoid close to lissamphibians, according to Marcello Ruta's phylogeny. It is a very specialized species of amphibamid dating from the Late Carboniferous. Found

Doleserpeton

Age Early Permian – 285 Ma
Location United States
Size Decimetric
Features Rounded skull; large orbits; pedicelate and bicuspid teeth
Classification Amphibamid dissorophoid temnospondyl

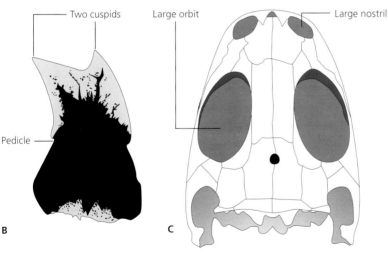

- Two cuspids
- Pedicle
- Large orbit
- Large nostril

B   C

Amphibamus

Age Late Carboniferous
Location United States
Size Up to 10 cm
Features Cylindrical pleurocentrum; light skeleton; large orbits
Classification Amphibamid dissorophoid temnospondyl

at the famous fossil site of Mazon Creek, Illinois, this small temnospondyl is characterized by a cylindrical pleurocentrum and a very light skeleton, characters which draw it closer to the lissamphibians. A carnivore and probably terrestrial or amphibious (its lifestyle is still debated), it probably lived near freshwater.

However attractive these hypotheses may be, we should not forget that they are not the only ones on the table: some paleontologists place the sister group of lissamphibians in another group of stegocephalians, the lepospondyls; others even contest their monophyly.

The Pangaean Chronicles

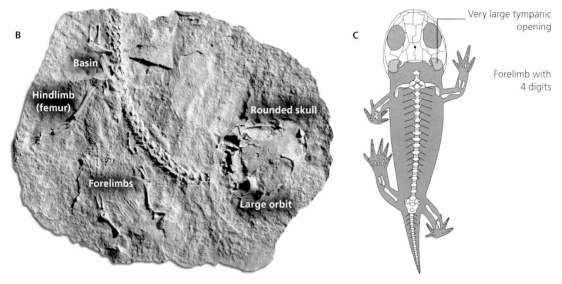

3.10. (A) *Amphibamus grandiceps* (Late Carboniferous), the closest relative of lissamphibians according to certain phylogenies. The controversy surrounding the origin of modern amphibians may never be solved as long as paleontologists lack information on the growth of these fossil species. (B) Cast of the fossil; (C), reconstruction of its skeleton.

Photo reproduced with the kind permission of John Bolt (Field Museum of Natural History, Chicago).

Back to the Roots: *Edingerella madagascariensis*

Most temnospondyls we have met were, as far as we can judge, terrestrial or amphibious. There were other representatives, however, that lived in an aquatic environment. The surprising thing is that these forms underwent a *secondary readaptation* to an aquatic way of life – in freshwater, coastal zones, or even open seas.

Let us look more closely at these marine forms, which we can say went "back to their roots" because their Devonian ancestors were mostly marine (see chapter 1). The return to the sea is visible from the Late Carboniferous–Early Permian (300 Ma) in, for example, *Iberospondylus*, which was a coastal form. From the Early Triassic (250 Ma), marine temnospondyls become increasingly numerous and truly pelagic, and

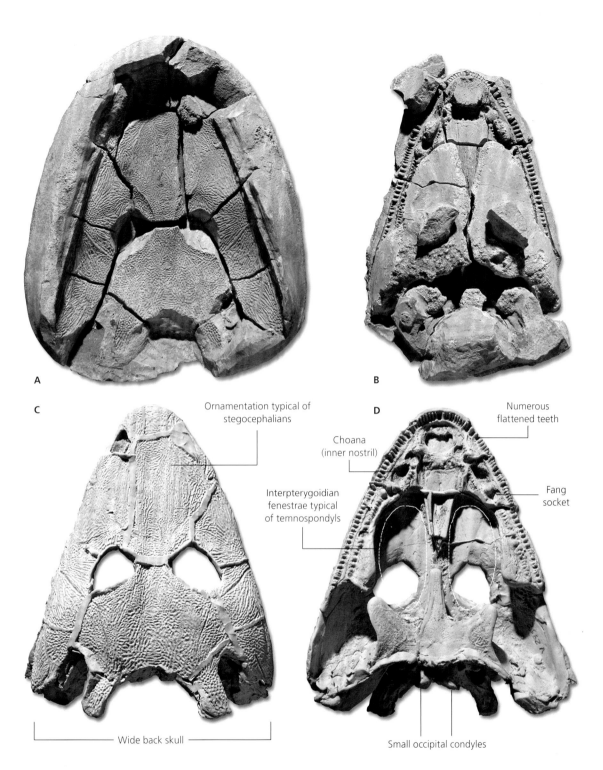

3.11. A marine temnospondyl: *Edingerella madagascariensis* (Early Triassic of Madagascar): skull counterpart preserved in a nodule (A: dorsal part; B: palatal part); mold of this counterpart realized in polymer resin (C: dorsal view; D: palatal view). The occipital condyles correspond to the attachment zone of the skull into the vertebral column.

Photos reproduced with the kind permission of Simone Maganuco (Natural History Museum of Milan, Italy).

3.12. *Edingerella madagascariensis* (bottom right) watched by a hybodont shark (above). *E. madagascariensis* is a Triassic temnospondyl that readapted to a marine way of life. Phylogenies suggest that its ancestors were terrestrial. How can we talk about *one* terrestrialization?

Edingerella

Age Early Triassic–250 Ma
Location Madagascar
Size Up to 1m
Features Rather long snout; dermal sensory canals on skull
Classification Watsonisuchian capitosaur temnospondyl

hydrodynamic forms like some trematosaurs, a group of piscivorous temnospondyls with long snouts, appear. We will look at the example of *Edingerella madagascariensis* (figs. 3.11 and 3.12), discovered in Madagascar several decades ago in deposits dating from the Early Triassic.

This marine temnospondyl was the focus of Simone Maganuco. For his dissertation, he redescribed all the specimens put at his disposal by the natural history museums of Milan (Italy) and Paris (France). All had been found in the north of Madagascar—together with fish, crustaceans, and marine cephalopod mollusks (ceratites). This associated fauna indicates a rather pelagic (open-sea) environment. What was *Edingerella*'s morphology? Of modest size, adults reached barely a meter in length; this temnospondyl possessed a long, wide, and flattened snout and dermal sensory canals running the length of its head. These must have allowed it to escape big predators and to detect small prey (such as fish or other amphibians).

Edingerella and the other marine temnospondyls had to confront the same problem as Devonian marine tetrapods: how do you tolerate saltwater when you are an amphibian? As we saw in chapter 1, this problem of euryhalinity has not yet been resolved. What adaptive strategies did these temnospondyls develop? Perhaps they had a salt gland, an organ which manages osmotic stress and can be seen today, particularly in marine turtles and crocodiles and the pinnipeds (which include seals and walruses).

It is highly probable that the readaptation of temnospondyls to a marine environment favored their rapid dispersion around the globe from the end of the Carboniferous to the beginning of the Permian.

The "Precious" of Lesotho: One of the Largest Amphibians of All Time

At the beginning of the 1970s, during an expedition to the heart of Lesotho in southern Africa, a team of paleontologists (Bernard Battail, Paul Ellenberger, and Léonard Ginsburg) discovered the skull fragment of a strange temnospondyl in deposits dating from the Late Triassic–Early Jurassic (210 Ma). The 20-centimeter fragment was carefully packed and brought back to the Muséum National d'Histoire Naturelle in Paris. In the depths of the museum laboratory, the fossil found its way into the hands of preparator Philippe Richir. Once cleaned and freed from its matrix, it revealed an unexpected form that set tongues wagging. Based on its ornamentation and the shape of its teeth, it was thought to belong to a large temnospondyl from the capitosaur group (to which *Edingerella* also belongs).

The piece came back to one of its discoverers, Léonard Ginsburg, who published on it with a colleague and then hid his "precious" away from prying eyes at the back of a drawer. And some things that should not have been forgotten were lost. Years went by. The specimen passed out of all knowledge. History became legend. Legend became myth – that of the only temnospondyl from Lesotho. For nearly thirty years no one talked of the capitosaur from Lesotho until, when the chance came, the fossil ensnared another bearer, a paleontologist called Philippe Janvier. For years, lurking in the unlit interior of a drawer, it occupied his mind. Darkness crept back into the Jardin des Plantes. Rumors grew of a shadow – whispers of a nameless fear. The Lesotho fossil knew its time had come. But then something happened the fossil did not expect. Philippe gave me the fossil and said, "Léonard's fossil . . . He's left to work on mammals. I've kept it all these years and now it's yours. Re-examine it carefully and, above all, keep it safe. I have a few matters I must attend to."

So now I had the "precious" in my hands and the task of redescribing it in light of contemporary knowledge. It was indeed a snout fragment from a large temnospondyl, as my predecessors had shown, but was it that of a capitosaur? It looked doubtful because the specimen was very curved in profile, whereas capitosaurs have flat snouts. Moreover, the teeth, when examined under the microscope, displayed a labyrinthodont-like folded structure, but the degree of folding of the dentine suggested that I was not dealing with a capitosaur but a brachyopoid, another group of temnospondyls. This fine distinction had important repercussions. Rapid calculation based on size proportions of the best-known and complete representative of this group indicated that the fragment belonged to a

3.13. One of the largest amphibians of all time. (A) The "precious" of Lesotho is a 210-million-year-old skull fragment. (B) It belonged to a brachyopoid, a temnospondyl that reached up to 8 meters in length! This fossil is not complete enough to erect a new taxon. We can only assign it to a family.

Photo reproduced with the kind permission of the Muséum National d'Histoire Naturelle/Denis Serrette.

skull around 2 meters wide from an animal measuring nearly 8 meters in length! Although incomplete, the "precious of Lesotho," was probably even bigger than *Mastodonsaurus giganteus*, which measured around 6 meters, from the Triassic of Germany. After decades of silence, this "precious" had finally given up its secret: I was looking at one of the largest amphibians the Earth has ever known (fig. 3.13).

Sahara before the Dinosaurs: At the Heart of Pangaea during the Late Permian

Niger is a place dear to paleontologists, and is one of the few spots on the planet (along with Russia and South Africa) that kindles the hope that one day we can reconstitute the flora and fauna of the Permian world. In this African country are precious exposures of continental sediments dating from this period. Perhaps even more exciting is that, in the Late Permian, the Niger that we know today was located at the heart of Pangaea (the supercontinent which fused all present-day continents into one; see fig. 3.23). Niger's fossils furnish us with valuable clues about tetrapods that inhabited the continental landmass at that time.

The first were discovered in the 1960s when, in the heart of the Sahara, geologists and mining prospectors stumbled upon almost complete dinosaur skeletons! Hearing of these finds, Jean-Pierre Lehman, then director of the paleontological lab at the Muséum National d'Histoire Naturelle, sent a young student called Philippe Taquet out to investigate. The scale of the discovery was so great that this young student did his dissertation on the Cretaceous dinosaurs of Niger. He also took the opportunity, together with his colleague Armand de Ricqlès, to prospect the Permian exposures in the north of the country. They brought to light a good portion of a large herbivorous terrestrial reptile, the skull of which was armed with an impressive battery of teeth: the famous *Moradisaurus grandis* (figs. 3.14 and 3.15). They published a description of its the skull in 1982. At that time it was the only published Permian tetrapod from the center of Pangaea. Others had come from "reference sites" located on the edges of the supercontinent—that is to say, the great sedimentary basins

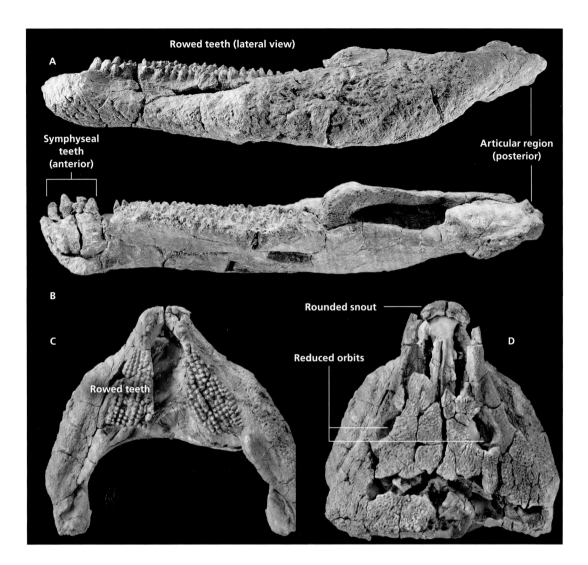

3.14. *Moradisaurus grandis*, a large herbivorous reptile discovered in the Niger desert (Moradi Permian deposits): (A) mandible in external view (labial), (B) internal (lingual) and (C) dorsal; (D) skull in dorsal view.

Photos reproduced with the kind permission of the Muséum National d'Histoire Naturelle/Denis Serrette.

of Russia and the Karoo Basin of South Africa. These latter have yielded relatively complete and homogenous faunas, principally made up of synapsids ("mammal-like reptiles") such as the herbivorous dicynodonts and carnivorous gorgonopsians.

In their description of the *Moradisaurus* skull, Taquet and de Ricqlès mentioned poorly preserved remains of other tetrapods. I pricked up my ears, as did Christian Sidor of the University of Washington. Perhaps tetrapods other than reptiles could be found at the heart of Pangaea 250 million years ago, just before the great Permian–Triassic life crisis. Temnospondyls? A prospection had to be carried out in the dry, flat region of Niger (fig. 3.16); an area somewhat inaccessible but where Permian outcrops make up the whole lunar—or, rather, Martian—landscape!

After getting together the funding (thanks to the National Geographic Society), we arrived in 2003. Our team, international and multidisciplinary, was composed of French, American, and Canadian

3.15. An inhabitant of central Pangaea in the Late Permian: *Moradisaurus grandis* (a herbivorous reptile).

Moradisaurus

Age Late Permian
Location Niger
Size Up to 2 m
Features Triangular skull; battery of teeth; robust limbs
Classification Captorhinid reptile

3.16. The Late Permian outcrops of Niger. This vast African country, traversed by the Sahara in the north, is very rich and diverse from a geological point of view. The zone in red was prospected during two international scientific expeditions in 2003 and 2006. The returns were excellent (see following page).

paleontologists; a South African geologist; and Nigerien archeologists—not to mention our Touareg guide, a doctor, and three soldiers from the Nigerien army to watch over us. We would be spending our time in the extreme north of the country, near the Algerian border, a region of "persistent insecurity" (the term officially used by the Nigerien government) due mainly to rampant trafficking. For more than a month we surveyed a hostile environment (fig. 3.17), eyes riveted to the ground, looking for the slightest clue, the tiniest bone splinter, walkie-talkies in hand to alert the rest of the group in the event of a find (fig. 3.18). Temperatures hovered around 50°C–requiring an individual, depending on build, to drink on average about 10 liters of water per day.

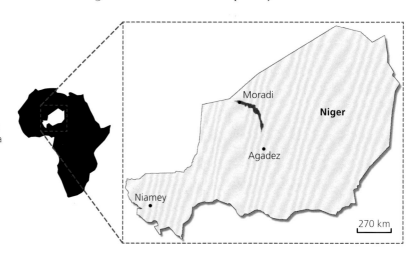

3.17. (A) The Moradi Formation in the Permian Red Stone Desert (Reg) of the Saharan Niger. (B) The surface of Mars. Strange similarities between the two desert landscapes: one scoured by an international group of researchers looking for fossils: the other by space probes that allow us a glimpse of some regions of the red planet.

Photos: Sébastien Steyer (CNRS/Muséum National d'Histoire Naturelle; A); NASA/JPL/Cornell University B).

3.18. Looking for temnospondyls in the heart of the Niger desert. Under a blazing sun and temperatures of 50°C, are Roger Malcom Smith (right), geologist and stratigraphist at the Iziko Museum of South Africa, Cape Town, Robin O'Keefe (left background), paleontologist at the Marshall University (West Virginia), and yours truly (left foreground).

Photo reproduced with the kind permission of Christian Sidor (University of Washington).

The Pangaean Chronicles   85

3.19. *Nigerpeton,* one of the surprising stegocephalians from the Late Permian discovered in the Niger desert in 2003: (A) left portion of the skull; (B) reconstruction sculpted by Franck Limon-Duparcmeur.

Photos: Christian Sidor (University of Washington; A); Franck Limon-Duparcmeur B).

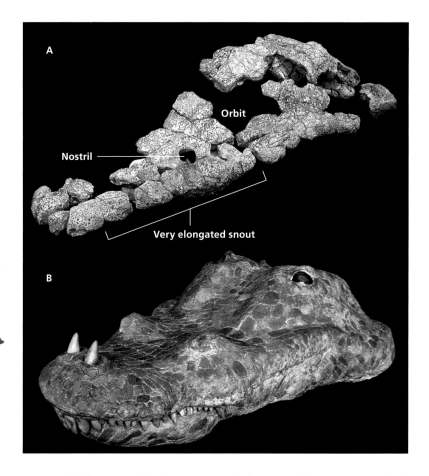

Nigerpeton

Age Late Permian
Location Niger
Size Up to 3 m
Features Very long skull; natural opening for teeth perforation at end of snout
Classification Cochleosaurian edopoid temnospondyl

3.20. (facing) Close up of *Nigerpeton:* skull in (A) dorsal, (B) palatal, and (C) lateral right views; and (D) reconstruction. This carnivorous temnospondyl was probably more aquatic than *Saharastega* (see following page). *Nigerpeton* would have lived in temporary rivers in central Pangaea during the Late Permian. The fangs of its lower jaw naturally pierce the upper jaw, as observed in some crocodiles today. Equipped with a very long snout and orbits positioned on the top of the skull, this 2-meter-long temnospondyl must have occupied the same ecological niches as some crocodile species do today.

It took about a week of prospecting before our efforts were rewarded. First we found remains of *Moradisaurus,* the large herbivorous reptile that, until then, was known only from its skull. Next came temnospondyls! They belonged to different genera: one we baptized *Nigerpeton,* "the fossil amphibian of Niger" (figs. 3.19 and 3.20); the other *Saharastega,* "the fossil amphibian of the Sahara" (fig. 3.21).

These temnospondyls, with their surprising morphologies, are completely new and, for the moment, the only existing specimens. They must have lived in pools or oxbows. Their osteological description and the reconstitution of their paleoenvironment appeared in an article published in *Nature* in 2005. These two amphibians are not close in the Nigerien family; they occupy two different positions in the temnospondyl phylogenetic tree (fig. 3.22). The closest taxa to *Nigerpeton* and *Saharastega* are temnospondyls 30 million years older, found in Carboniferous deposits situated in the north of Pangaea (Euramerica—present-day North America and Europe combined).

This strongly suggests that individuals of each lineage that led to our Nigerien temnospondyls traveled, over several millions of years, thousands of kilometers south into Gondwana (a vast landmass comprising Africa, South America, Australia, India, and Antarctica). How did these aquatic or amphibious organisms make such a long migration?

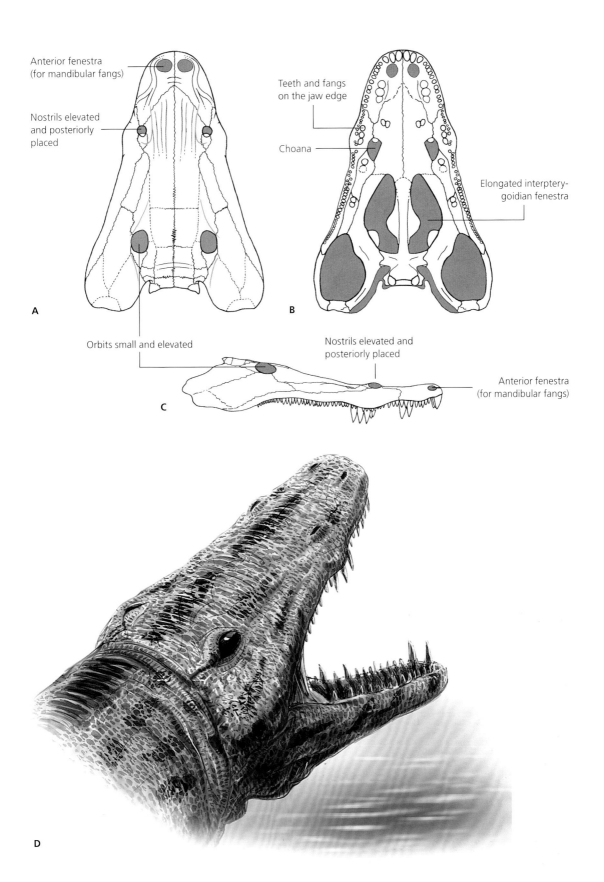

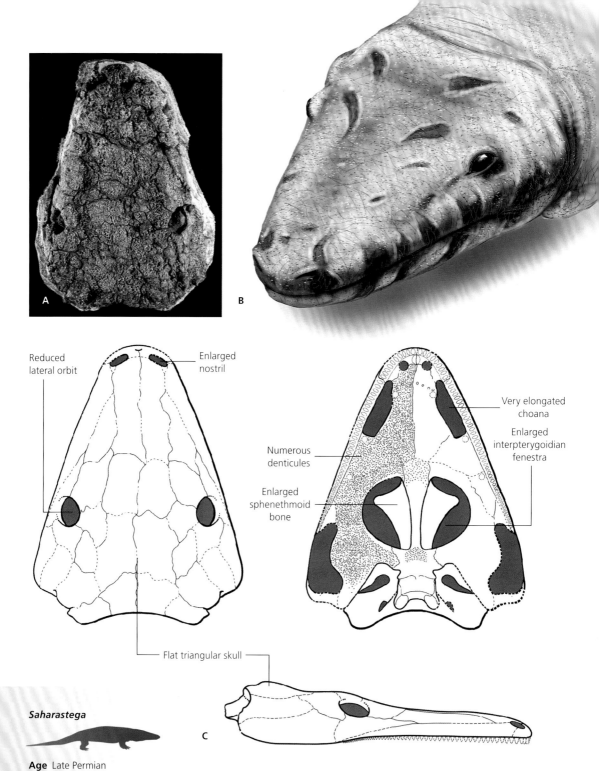

Saharastega

Age Late Permian
Location Niger
Size Up to 1.5 m
Features Triangular skull; small lateral orbits
Classification Temnospondyl close to edopoids

3.21. *Saharastega*, another central Pangaean temnospondyl from the Late Permian: (A) dorsal view of the skull as preserved; (B) reconstruction; (C) skull in dorsal, palatal, and lateral right views. *Saharastega* possessed a markedly triangular skull with small lateral orbits. Its palate was armed with a multitude of small teeth that carpeted the upper buccal cavity. This odd carnivorous stegocephalian grew up to 1.5 meters.

Photo reproduced with the kind permission of Christian Sidor (University of Washington).

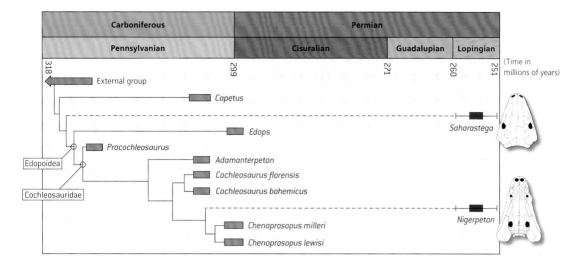

3.22. Phylogenetic relationships between the Nigerien temnospondyls *Nigerpeton* and *Saharastega* (in red), and Euramerican relatives (in blue). This cladogram clearly shows that the two Nigerien taxa do not constitute an African clade. They are related to much older northern Pangaean forms from Euramerica. The populations that formed the lineages of *Nigerpeton* and *Saharastega* had to undertake long journeys into southern Pangaea or Gondwana (to which present-day Niger belongs).

3.23. Pangaea during the Late Permian, with possible migration routes used by close relatives of the Nigerien temnospondyls *Saharastega* and *Nigerpeton*. From populations in the north of Pangaea (Euramerica), two lineages successfully migrated towards Gondwana. They could have taken two possible routes: an intracontinental route cutting through the Hercynian Chain (fluvial route, in orange), or a maritime route along the coasts (coastal route, in blue). Each lineage perhaps chose its own route.

The Pangaean Chronicles   89

3.24. Everyday life in the heart of Pangaea, 250 Ma. Left, *Bunostegos akokanensis*, a pareiasaurian (large herbivorous reptile), the skull of which displays surprising lumps (photo left). To the right, two gorgonopsians lurk in the shade. The tooth (photo right) is one of the rare remains we found of these large carnivorous synapsids from the Late Permian Nigerien deposits.

Bunostegos

Age Late Permian
Location Niger
Size Up to 3 m
Features Bony bosses on skull; robust limbs
Classification Pareiasaurian reptile

Two non–mutually exclusive scenarios are plausible (fig. 3.23). The fluvial route scenario suggests that they used the extensive watershed network (streams, rivers, and other watercourses) that drained the Hercynian Chain, a mountain range that separated Pangaea into Laurasia to the north and Gondwana to the south. This would not have presented an insurmountable obstacle to these amphibians since, according to geologists, it was already highly eroded, with a maximum elevation of 2000 meters. In the coastal route scenario, the relatives of *Nigerpeton* and *Saharastega* took the coastal route bordering Pangaea, thus passing from the north to the center, then toward the interior of Pangaea: perhaps,

Photos reproduced with the kind permission of Christian Sidor (University of Washington; left); Sébastien Steyer (CNRS/Muséum National d'Histoire Naturelle; right).

like all pioneering marine explorers, these temnospondyls were master navigators! The coastal scenario implies a much longer migratory route and saltwater tolerance (euryhalinity), seen also in temnospondyls that returned to a marine existence at the end of the Permian and the beginning of the Triassic (see pp. 78–80).

Whatever the truth of the matter, our Nigerien expedition proved particularly rich in finds, but not solely limited to amphibians. *Nigerpeton* and *Saharastega* form two components of a fairly complete and endemic fauna also composed of reptiles. Apart from *Moradisaurus* mentioned earlier, we discovered several near-complete skeletons of a new pareiasaurian.

Gorgonopsian

Age Late Permian
Location Niger (and Africa in general)
Size Plurimetric
Features Long shearing canines
Classification Non-mammalian synapsid ("mammal-like reptile")

The Pangaean Chronicles

This terrestrial herbivorous reptile–solid, thickset, and several meters long–had a surprising morphology: its skull showed natural lumps or bosses (which may have supported horns), and its whole body was armored. We named it *Bunostegos akokanensis*, "the lump-headed reptile of Akokan," the locality nearest the fossiliferous site (fig. 3.24). Another unusual reptile, smaller but also herbivorous, was found during our stay. Quite similar to *Moradisaurus*, it belonged to a group of captorhinids, some representatives of which are characterized by an odd downward-curved snout.

From the indications the Niger desert provides us, the central Pangaean fauna of the Late Permian was mostly composed of herbivorous reptiles and carnivorous amphibian temnospondyls. The presence of large carnivorous synapsids ("mammal-like reptiles") called gorgonopsians is suggested by large canines and a partial jaw we found in the Moradi Formation (fig. 3.24). But most of the carnivores of this Nigerien fauna are amphibians, in contrast to other Permian faunas on the edges of the supercontinent, where the dominant carnivores are the amniotes. In addition, let us not forget that the species discovered in Niger are new–that is to say (and until proved otherwise), endemic.

The site we explored is, therefore, very interesting, as it casts doubt as to the homogeneity of Pangaean faunas. Furthermore, faunal homogeneity is one of the arguments cited to suggest a uniform hot and dry climate over the entire supercontinent. The temnospondyl amphibians and the reptiles discovered, as well as analyses of the sediments that yielded these fossils, suggest that in the Late Permian, the climate in fact was variable at the heart of Pangaea, markedly alternating between wet and dry seasons. Stratigraphical analyses confirm this hypothesis. In order to glean the maximum paleoclimatic information, we returned to the area in 2006 with paleobotanists and geochemists. This expedition was also very fruitful: xerophile plant remains, silicified tree trunks several meters long, tree stumps in situ, and leaves were all found in the Moradi Formation. Paleosoils were also discovered, samples of which have been taken for geochemical analysis–an ongoing process. All these studies will soon allow us to get a pretty good idea of the paleolandscape of the period.

In the Shadow of Temnospondyls: The Lepospondyls

Next to the abundant and highly diversified group that formed the temnospondyls, other more discrete amphibians developed from the end of the Paleozoic. These were the lepospondyls already encountered in this chapter. These batrachomorph stegocephalians of generally modest proportions developed from the Carboniferous to the Permian in Euramerica, with only one Gondwanian exception. They did not experience as large an expansion, neither temporally nor geographically, as their temnospondylian cousins. The lepospondyls seem to have lived in the shadow of their cousins, who were of a more imposing size and probably above them in the food chain, at least as adults.

Anatomically, the lepospondyls are very interesting: they are characterized by bobbin-shaped vertebrae and their teeth do not have a labyrinthodont structure, unlike other stegocephalians. They form an entirely separate group, generally defined by reduced limbs and a lengthening of the body—lending them an eel-like appearance in some cases, and salamandriform in others. Some developed highly original morphologies, as we shall see. Mostly found in sediments suggesting continental environments (such as lakes and swamps), these amphibians are principally terrestrial, amphibious, or aquatic, just like the temnospondyls. There is a difference in that no lepospondyl has, at present, been found in typical marine sediments.

We now look at some choice morsels from the lepospondyl group, starting with the morphological curiosity that is *Diplocaulus*, and moving on to the aistopods, which lost their limbs but may have been among the earliest tetrapods to venture onto land. We will finish off with the lysorophoids, second on the shortlist for the closest relative of the lissamphibians.

Boomerang Head: *Diplocaulus*

After *Spathicephalus* and its Frisbee skull comes *Diplocaulus*, a lepospondyl with a boomerang-shaped head, discovered in the Permian of North America (fig. 3.25). It looks decidedly as if the stegocephalians were champions of innovation, pushing the limits of morphological possibility.

Diplocaulus possessed rather reduced limbs as well as long, straight, fine ribs that lent its body a very flattened appearance. What is most striking, however, is the skull shape. But *Diplocaulus* is not the only vertebrate with a boomerang head: *Zenaspis pagei* (an ostracoderm from the Scottish Devonian), *Sphyrna* sp. (the present-day hammerhead shark), and *Gerrothorax pulcherrimus* (a plagiosaurian temnospondyl from the German Triassic) also have skulls that conjure up images of UFOs (fig. 3.26). Here we have a fine example of convergent evolution, even though each "boomerang" has its own particular form. For example, in *Sphyrna*, the orbits are located on the ends of the "horns." This morphology is linked to hydrodynamism, of course, but also available surface is increased to develop sensorial organs on the head. In the other three organisms, which are fossils, the orbits are situated in the center of the skull, in a dorsal position. This is often the case in benthic species today.

Diplocaulus, a freshwater stegocephalian, probably favored the bottoms of watercourses, even the muddy beds, where we can imagine it lying camouflaged with only its small globular eyes visible. When it moved around, its short limbs were probably tucked close to its body.

A question that naturally springs to mind is whether some sort of selective advantage was to be had from its boomerang head. Its unusual form may not have any particular function: not every form is necessarily associated with a given function (the reverse is also true). This does not

Diplocaulus

Age Late Permian
Location United States and Morocco
Size Up to 1 m
Features Flat, boomerang-shaped skull
Classification Diplocaulid lepospondyl

3.25. (A) A group of *Diplocaulus* in freshwater. With their "boomerang heads," these Permian lepospondyls are amongst the best-known fossil amphibians. (B) The well-preserved skull of a diplocaulid discovered in Morocco. As is the case with all the representatives of this group, it is extremely flat, triangular, and shows two long lateral horns, the function(s) of which are still hotly debated.

Photo reproduced with the kind permission of the Muséum National d'Histoire Naturelle/Damien Germain.

prevent us from putting forward hypotheses. These horns could have served to seduce females or to intimidate males – or predators. The very wide skull could also have improved lift while swimming, allowing it, as the evolutionist Stephen Jay Gould proposed, to swim faster against the current when hunting. Debate continues about this lepospondyl that excites our imagination.

Diplocaulus belongs to the family Diplocaulidae, all from Permian deposits in the northern part of Pangaea. There are exceptions, however, including *Diplocaulus* itself. In 1970 paleontologist Jean-Michel Dutuit discovered in Morocco (formerly part of Gondwana) remains that he attributed to this genus (they have recently been re-examined). This important find made *Diplocaulus* the only lepospondyl whose traces have been found in Gondwana. Recently, with Nour-Eddine Jalil (University of Marrakech), Renaud Vacant (CNRS, Paris) and Damien Germain (Muséum National d'Histoire Naturelle, Paris), we followed in the footsteps of Dutuit, back to the sedimentary basin of Argana (in the western Atlas). We unearthed more remains of *Diplocaulus*, which have recently been described together with the original material (fig. 3.27). The results have brought about important revisions concerning the distribution and morphology of this amazing lepospondyl which migrated "out of Laurasia" and possessed a slightly asymmetrical boomerang head!

Fake "Micro-Reptiles": The Microsaurs

In the humid, lake-dotted landscape of equatorial Pangaea, discreet, small lepospondyls called microsaurs rubbed shoulders with the flamboyant diplocaulids. Known from the Early–Late Carboniferous to the late

3.26. The "Boomerang Club." Vertebrate skulls have converged toward a boomerang shape several times during evolution. The shape provides lift to the head moving through water (or mud) as it does a boomerang thrown into the air. Shown here: *Diplocaulus magnicornis*, *Zenaspis pagei* (an ostracoderm from the Devonian of Scotland), *Sphyrna* sp. (a present-day hammerhead shark), and *Gerrothorax pulcherrimus* (a plagiosaurian temnospondyl from the Triassic of Germany).

3.27. (following page) On the trail of *Diplocaulus* in the Moroccan Atlas. Surveys and excavations conducted by yours truly in the south of Morocco, notably (A) with N. E. Jalil (University of Marrakech); and (B) R. Vacant (CNRS, Paris), have led to the discovery of new *Diplocaulus* remains, including a vertebra (C) and ribs (D).

Photos: Sébastien Steyer (CNRS/Muséum National d'Histoire Naturelle).

96   Earth before the Dinosaurs

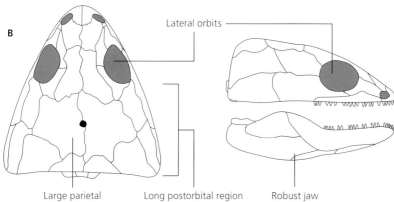

3.28. (A) *Microbrachis*, a Late Carboniferous microsaur. This lepospondyl conserved larval characters in its adult form, including gills. This phenomenon, known as neoteny, is common in amphibians. The axolotl, with external gills that persist after metamorphosis, is an example of neoteny in living amphibians. (B) The skull in dorsal and lateral right views.

Early Permian of Euramerica, they represent the most diversified group of lepospondyls, with more than 20 genera identified to date. They come in numerous forms, very often slender bodied with delicate limbs that suggest an amphibious or aquatic existence.

Their morphology reminds us of salamanders or small lizards. Previously considered "fossil reptiles" (hence the name), their ancestral links to other lepospondyls is still hotly debated. They have smaller orbits than other lepospondyls and the closed skull lacks large natural openings: microsaurs do not have an otic notch behind the skull, and their very large supratemporal separates the squamosal from the parietal. In addition, they have two occipital condyles that link up with the vertebral column.

Take, for example, *Microbrachis*, a microsaur of about 15 centimeters with quite a long body and a rounded skull (fig. 3.28). Found in the Late Carboniferous of Nyrany, a famous fossil site in the Czech Republic, this lepospondyl has more than 40 vertebrae (compared to between 15 and 26 in its close cousins). Apparently amphibious or aquatic, it would have

Microbrachis

Age Late Carboniferous
Location Czech Republic
Size Up to 15 cm
Features Long body (up to 40 vertebrae); short limbs; rounded skull
Classification Microsaurian lepospondyl

The Pangaean Chronicles   97

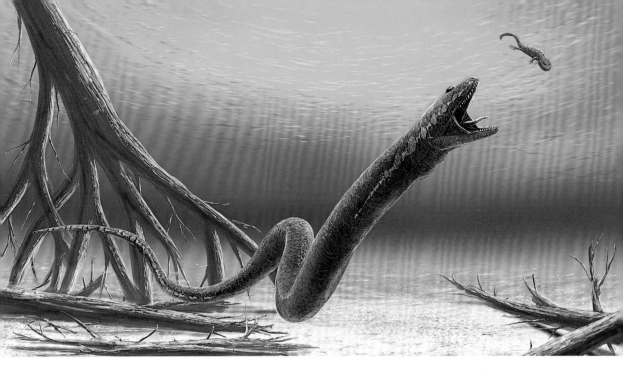

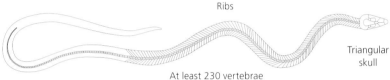

3.29. *Ophiderpeton*, aistopod from the Late Carboniferous of Europe and North America. Aistopods are the earliest tetrapods to have lost their limbs. Snakelike *Ophiderpeton*'s 70-centimeter body shows at least 230 vertebrae. A carnivore and probably a burrower, this lepospondyl likely fed on worms, arthropods, and even other small tetrapods. *Ophiderpeton* was discovered in palo-lacustrine sediments (deposited in swamps and lakes). Its lifestyle, amphibious or terrestrial, is still a matter of debate.

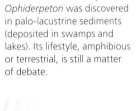

Ophiderpeton

Age Late Carboniferous
Location North America and Europe
Size Up to 70 cm
Features Snakelike body (more than 230 vertebrae) without limbs; triangular skull
Classification Aistopod lepospondyl

moved through the water by lateral undulation, with little help from its short limbs. It is quite possible also that it fed on freshwater zooplankton.

Limbless but Terrestrial? The Aistopods

Some lepospondyls, the aistopods, display the peculiarity of having lost their limbs. This is the first time in vertebrate evolution that a tetrapod group lost its limbs (it happened later to some little-known modern amphibians—the caecilians—and, of course, to snakes).

Present between the Early Carboniferous and the Early Permian, the aistopods are characterized notably by an enlargement of the postorbital and by very mobile snout bones. Could this cranial "plasticity" (not as marked as in snakes) have allowed aistopods to swallow prey of a considerable size? Studies are being carried out to answer this question. The lifestyle of aistopods is still, in most cases, a matter of dispute: Were these amphibians terrestrial or amphibious? In his Ph.D. dissertation at the Muséum National d'Histoire Naturelle, Damien Germain has put forward a provocative hypothesis to answer this question: he suggests that Early Carboniferous aistopods were preponderantly terrestrial. This would mean that, among the first terrestrial tetrapods, there were some that had already lost their limbs!

Aistopods had rather long bodies (fig. 3.29). In living snakes and caecilian amphibians it is interesting to note that reduction or loss of limbs is often accompanied by a multiplication of vertebrae and a lengthening of the trunk during morphogenesis. This is also true of aistopods, which are often snakelike. This would have been useful to undulate effortlessly from an aquatic to a terrestrial environment.

Let us end by noting that aistopod lepospondyls have been found in France in 300-million-year-old deposits, notably at the exceptional fossil site of Montceau-les-Mines (Saône-et-Loire).

The Origin Is Elsewhere? The Lysorophians

We return once again to the controversial question of lissamphibian origins. Some authors believe their sister taxon is to be found within the dissorophoid temnospondyls; others situate it within the lepospondyls and, more specifically, the lysorophians.

The lysorophians are doubtless the most enigmatic lepospondyls. Their lifestyle is a bone of contention, although they would have been mostly aquatic; S-shaped traces of crawl marks found on the fossilized beds of watercourses are attributed to these organisms. Furthermore, most lysorophians from the American Permian have been found in nodules that appear to be the remains of mud burrows built in pond beds. This would have been a strategy developed to survive drought and desiccation, much in the manner of dipnoi (see chapter 1).

Present from the beginning of the Late Carboniferous to the Early Permian in Euramerica, lysorophians were highly specialized and, therefore – as you will no longer be surprised to learn – are difficult to classify. They are characterized by a rather delicate skull that is long and fenestrated with large nostrils; large, extremely lateralized orbits; and four highly reduced limbs of equal proportion (fig. 3.30). The scapular and pelvic girdles are not well ossified, and their neural arches fuse with the vertebral elements. The vertebral column is extremely long. This lengthening of the trunk, already observed in several lepospondyl groups, gave them an almost snakelike appearance. It also appears that the body of lysorophians lengthened during their evolution: the trunks of Late Carboniferous representatives possess 69 vertebrae, compared to 97 in Early Permian forms.

Of all these morphological characters, which bring the lysorophians and lissamphibians closer together, thereby suggesting a lepospondyl origin for modern amphibians? We can reproach advocates of such a phylogeny for having founded their analysis on a large number of unknowns or character reversions, such as the absence of a snout and of the pineal foramen, as well as missing skull bones (postfrontal, postorbital, jugal, and ectopterygoid, for example). Once again we see the issue is far from resolved and I am tempted to shelve these phylogenetic debates until we find, for example, fossil "proto-lissamphibians" in Paleozoic deposits. It is better to spend the time in the field than in front of a computer running endless phylogenetic simulations that are of little use without more field data.

Brachydectes

Age Early Permian
Location United States
Size Up to 15 cm
Features Very long body; very reduced limbs and girdles
Classification Lysorophian lepospondyl

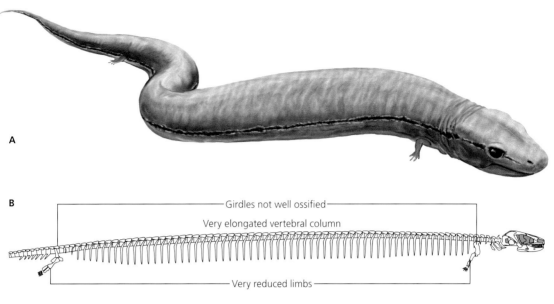

3.30. (A) *Brachydectes,* a lysorophian lepospondyl from the Permian of the United States. According to Laurin and Reisz (1997), lysorophians are at the origin of all lissamphibians. According to others, they are already too specialized to be their sister taxon. Feelings run high on this issue. (B) Complete skeleton and skull in lateral right view.

Lissamphibians on the Edge

The lissamphibians are often described as modern amphibians, but they also have a long history that is still far from clear. They represent the only living non-amniotic tetrapods.

The lissamphibians are distinguished less by their size, which is quite modest (pluricentimetric), than by their variety. They can be swimmers, walkers, and even jumpers. Most are amphibious but some forms (such as the axolotl, a salamander that retains its external gills in the adult stage) are exclusively aquatic, whereas others, surprisingly, are terrestrial (such as desert frogs) or even flyers (such as tree frogs; see chapter 4). Their morphology is highly specialized: frogs and toads, excellent jumpers, have skeletons with no ribs, fused limb bones and ultra-light skulls – unique features among vertebrates. Lissamphibians also have a very special physiology: they metamorphose and have numerous cutaneous glands. In short, all these specializations render their classification a delicate task.

Lissamphibians are preponderantly freshwater forms. In fact most of them have lost the capacity their distant Devonian ancestors had to tolerate salty or brackish waters. In the past no one suspected this capacity even existed, because *Ichthyostega* and company were believed to be freshwater animals, like lissamphibians (see chapter 1). Modern saltwater amphibians are thin on the ground, to such an extent that during his voyage on the *Beagle*, Darwin himself was astonished at a Patagonian frog (*Pleuroderma bufonina*) "reproducing and living in waters too salted to be drunk."

Lissamphibians are ectotherms (their principal heat source is external) and, therefore, poikilotherms (their body temperature is dependent on the external environment). They are characterized by pedicellate teeth (articulated at the base by a small peduncle fixed into the jaw), two types of cutaneous glands (mucous and granular), and ribs which, when present, are not ventrally attached. They are made up of three orders still represented in the living world—the anurans (frogs and toads), the urodeles (salamanders and newts), and the caecilians (or gymnophiones)—to which must be added fossil representatives of contested affinity such as the albanerpetontids, difficult to place within the group's phylogenetic tree. It should be noted that relationships among the three living orders vary depending on whether molecular or anatomical data is used: we therefore do not know with certitude which are more closely related to the urodeles, anurans, or caecilians. However, nearly all specialists, barring the odd voice of dissent (see fig. 3.1), agree that lissamphibians form a clade—that is to say, a group with a common and unique ancestor. The real sticking point, as you are now aware, is the origin of the group: Is it to be found among the temnospondyls or the lepospondyls?

We now explore the fossil diversity of each lissamphibian group and get acquainted with the oldest known fossil anurans, urodeles, and caecilians.

Double Record: Triadobatrachus massinoti

Triadobatrachus massinoti is a valuable fossil that is a double record holder. Found in the Early Triassic of Madagascar and around 250 million years old, it is the oldest known representative of the lissamphibians and the oldest known anuran in the fossil record.

This tetrapod already possesses nearly all the attributes of a modern frog (fig. 3.31): a light and fenestrated skull, a frontal bone almost completely fused to the parietal, a toothless dentary (mandible), an ilium (pelvic bone) oriented forward, few presacral vertebrae, and no ribs. Nevertheless, the hindlimbs of *Triadobatrachus* remain relatively short and are composed of bones not yet fused together. Living anurans have very long hindlimbs that are used for jumping and constituted of fused bones. This Triassic form was not a jumper, unlike living frogs and toads. In addition, it still had a little tail, which was to disappear in its descendants. It is interesting to note that anuran tadpoles have a tail that they lose during metamorphosis through apoptosis (cell death).

Triadobatrachus is one of the most important fossils in the world: this key specimen, which illustrates the very early lissamphibians, is carefully

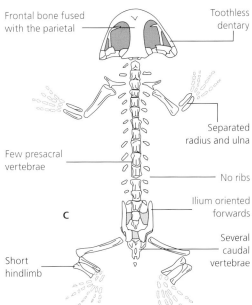

3.31. (A) *Triadobatrachus massinoti,* 250 Ma. It is both the oldest anuran and the oldest lissamphibian in the world – a double record! This fossil is known only from one specimen (B), counterparts of the skeleton preserved in a nodule). It has recently been restudied using microtomography (X-ray; see chapter 5). These analyses will perhaps bring us new anatomical data. (C) Reconstruction of the skeleton (characters of the genus in black, characters of anurans in red).

Photo reproduced with the kind permission of the Muséum National d'Histoire Naturelle/Denis Serrette.

Triadobatrachus

Age Early Triassic – 250 Ma
Location Madagascar
Size Up to 11 cm
Features Highly fenestrated skull; small tail; limbs of equal length
Classification Anuran lissamphibian

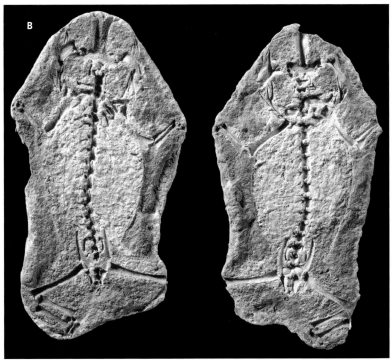

102   *Earth before the Dinosaurs*

housed in a safe at the Muséum National d'Histoire Naturelle. It consists of a ferruginous nodule resembling a pebble, with the fossil in counterpart—which means that the bones of the animal are not preserved but have left a faithful hollow print in the stone. All that had to be done to recover the exact form of the skeletal elements was to make a plastic resin mold. Luckily the skeletal imprint (or counterpart) of *Triadobatrachus* is almost complete. It appears that the nodule is of bacterial origin and developed around the cadaver of the animal in an aquatic environment. These slender clues point to *Triadobatrachus*'s having been amphibious like its living cousins, the frogs.

Described for the first time in 1936, this fossil was recently re-examined in detail by paleontologists Jean-Claude Rage (CNRS and Muséum National d'Histoire Naturelle) and Zbyněk Roček (Charles University and Science Academy, Prague). The verdict: despite being less of a jumper than present-day anurans, it already possessed highly specialized characteristics, opening up the possibility of an even more ancient origin for the lissamphibians. Who knows?—maybe one day we will discover another member of this group in Permian deposits.

Old Asiatic Salamanders: Chunerpeton and Karaurus

If the origin of anurans leans, for the moment at least, toward Africa, the history of the urodeles (or Caudata; salamanders and newts) seems to have started in Asia. The oldest known representative of this group is *Chunerpeton* (fig. 3.32), which was discovered recently in China, in Jurassic deposits dating from around 161 Ma (between the Middle and Late Jurassic). Equipped with a long tail, a highly fenestrated skull, and two sets of paired limbs almost identical in length (a typical urodele characteristic), it already resembles living salamanders.

Chunerpeton has even been attributed to a precise family of urodeles: the Cryptobranchidae, a family that notably includes *Andrias*, the giant Asiatic salamander (a living species; see inset "Sumo Salamanders"), and which is mostly made up of freshwater aquatic species. Was *Chunerpeton* also aquatic? The sedimentary and taphonomic context in which the specimens were found suggests it was. Unlike *Triadobatrachus*, known only from a single fossil, *Chunerpeton* is preserved in several hundred specimens. They were found in sediments corresponding to ancient volcanic ash apparently deposited on the surface of a lake, explaining their exceptional state of preservation: the gills (often external in aquatic urodeles), the ocular globes, and even the stomach contents have been fossilized! This will allow very detailed paleobiological analysis of this species. In addition, several growth stages of *Chunerpeton* are preserved, with individuals varying from several millimeters to around 20 centimeters in length. Comparison with living species of urodeles will hopefully shed some light on the origin and evolution of the group. These carnivorous and probably aquatic salamanders most likely fed on small shrimps (which were discovered in the same layers).

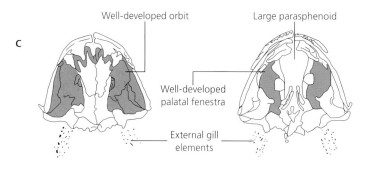

3.32. (A) *Chunerpeton tianyiensis* (Middle–Late Jurassic), the oldest known salamander to date. It belongs to the family Cryptobranchidae, as do present-day giant salamanders (see boxed text, pp. 108–109). Hundreds of specimens were recently found in China. They were probably quickly buried during a volcanic eruption 161 million years ago. (B) A beautifully preserved skeleton. (C) Skull in dorsal and palatal views.

Photo reproduced with the kind permission of Mick Ellison (American Museum of Natural History).

Chunerpeton

Age Middle–Late Jurassic—161 Ma
Location China
Size Up to 20 cm
Features Highly fenestrated skull; rather long body and tail
Classification Cryptobranchid urodele lissamphibian

Karaurus sharovi is another fossil representative of the urodele group that merits our attention. This specimen was discovered in Kazakhstan, in deposits dating from the Late Jurassic and described for the first time by the Russian paleontologist Ivachnenko in 1978 (fig. 3.33). *Karaurus*, a cast of which is housed at the Muséum National d'Histoire Naturelle, is also one of the oldest known salamanders. Slightly younger than *Chunerpeton*, its overall morphology is, however, less specialized than its Chinese cousin; it has a more voluminous, robust skull, showing dorsal ornamentation (small asperities and bumps). These cranial characters are plesiomorphic—that is to say, we encounter them in numerous stegocephalians other than urodeles.

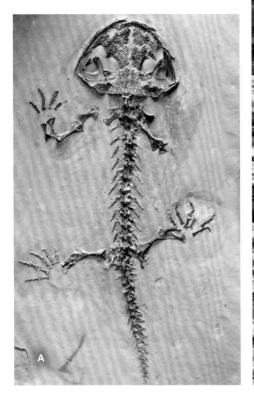

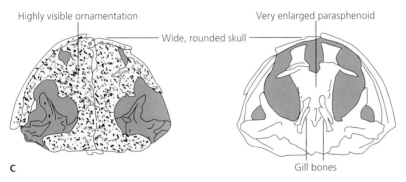

3.33. *Karaurus sharovi*, a salamander from the Late Jurassic of Kazakhstan. Slightly younger than *Chunerpeton*, its overall morphology is, however, less specialized. (A) An almost complete skeleton (dorsal view). (B) Reconstruction of the animal. (C) Skull in dorsal and palatal views.

Photo reproduced with the kind permission of Jason Anderson (University of Calgary); specimen from the Muséum National d'Histoire Naturelle.

First Caecilian with Limbs: *Eocaecilia micropodia*

The caecilians (or gymnophiones), the third and last group of lissamphibians, are without doubt the most enigmatic and intriguing of all. Their living representatives are more difficult to observe and rarer than other modern amphibians. Most are burrowers and found only in tropical and subequatorial latitudes. Hyper-specialized, caecilians possess a very long trunk composed of a multitude of vertebrae (up to 285!); and a small, cylindrical, non-fenestrated skull (lacking large openings) with greatly reduced lateral orbits, which appear from the larval stage and slightly resemble those of microsaurian lepospondyls (see pp. 95–97). But the most marked characteristic of living caecilians is the absence of limbs: this group lost its limbs during evolution.

Karaurus

Age Late Jurassic
Location Kazakhstan
Size Up to 20 cm
Features Rather robust and ornamented skull; stocky body
Classification Urodele lissamphibian

The Pangaean Chronicles 105

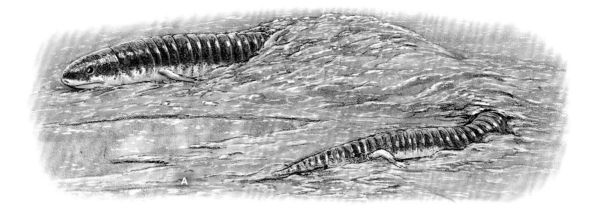

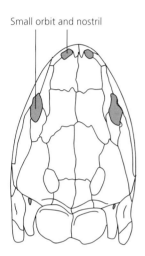

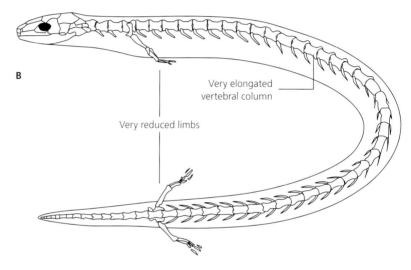

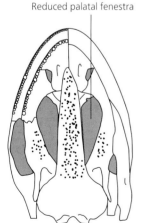

3.34. (A) *Eocaecilia micropodia,* the oldest known caecilian. If the caecilians of today are deprived of legs, as their name indicates, their oldest known representative (Early Jurassic of North America) still had limbs, albeit drastically reduced in size. Limb loss took place independently several times during tetrapod evolution (as we observe in aistopods and snakes, for example). (B) Skull in dorsal and palatal views, and skeleton.

This does not mean the caecilians originated from the aistopod lepospondyls (tetrapods that also lacked limbs; see p. 98). We must remember that limb loss occurred a number of times and in various groups throughout evolution. Specialists talk of *evolutionary convergence* when two taxa belonging to distant clades present the same traits, and of *parallelism* concerning two distinct taxa in closer clades. For example, limb loss corresponds to an evolutionary convergence between caecilians and snakes (lissamphibians and reptiles), and a parallelism between caecilians and aistopods (both amphibians).

The oldest known caecilian bears the name of *Eocaecilia* (fig. 3.34). Together with a very long trunk, it possessed limbs! Discovered in the Lower Jurassic of Arizona, this fossil is characterized by a relatively flat, non-fenestrated skull, a trunk composed of 49 vertebrae, and very reduced limbs (which were probably unable to support the animal's weight). It is also noteworthy that, unlike living caecilians, *Eocaecilia* was equipped with relatively well developed eyes: it was probably less of a burrower than other, later caecilians. After *Eocaecilia*, the oldest known caecilian is *Rubricacaecilia*. It was found in the Early Cretaceous (Berriasian) of Morocco: it is, therefore, also the first Gondwanian representative of the group.

Eocaecilia

Age Early Jurassic
Location Arizona, United States
Size Up to 14cm
Features Very long body (49 vertebrae); reduced limbs; non-fenestrated skull
Classification Caecilian lissamphibian

Threats of Yesterday and Tomorrow

As we have seen, the lissamphibians present a stunning diversity of forms. The oldest representative, the anuran *Triadobatrachus*, was a very specialized taxon as far back as the Early Triassic. Again, it could be that the evolutionary history of the lissamphibians dates back to the end of the Paleozoic. If this is the case, it raises the question as to how these smooth-skinned organisms, today so fragile and threatened with extinction, managed to survive the great Permian–Triassic life crisis that struck the Earth some 250 million years ago. One thing is certain: lissamphibians made it through the Cretaceous–Tertiary life crisis that brought about the extinction of the non-avian dinosaurs! The fossil record shows that the group even experienced a radiation just after this episode, at the beginning of the Paleogene. Could this apparent diversification be linked to a taphonomic phenomenon (more preserved sites than at other periods) or sampling (more specialists working on this period)? It is hard to say. Most of these small amphibious forms, whether aquatic or burrowing, perhaps lived in freshwater, lacustrine, or marshy environments—at least, that is how most live today—which were better protected and isolated from climatic changes. Or was the group opportunistic—that is to say, able to jump into ecological niches left vacant after the mass extinction, just as the mammals did after the Cretaceous–Tertiary crisis? Again, so many questions and so few answers.

The lissamphibians did pretty well throughout the Tertiary Period, but that is no longer the case: intensive deforestation and pollution are leading to the fragmentation or complete habitat disappearance of these very fragile species, which are sensitive to ultraviolet light (having no armored skin, scales, or hair). At present most lissamphibians are protected and yet still the group is undergoing a serious decline. Something must be done before it is too late: the world would be a sadder place if the amniotes (reptiles, birds, and mammals) were the only tetrapods to be found.

Sumo Salamanders

THE MUSÉUM NATIONAL D'HISTOIRE NATURELLE IN PARIS, stuffed with geological, mineralogical, paleontological and zoological samples, is recognized worldwide as a historical landmark. I am lucky enough to call it my workplace. Wandering its galleries is like traveling through time. Who can remain unmoved by the seemingly endless shelves of skeleton-filled jars? At the far end of the comparative anatomy gallery is the amphibian display case and, tucked away in a corner, is a truly remarkable skull: that of a giant salamander with the strange name of *Menopoma* scratched in India ink on an old, yellowing label.

Menopoma, a type of urodele lissamphibian, is the old denomination of the genus to which the giant American salamander belongs, today rebaptized *Cryptobranchus alleganiensis*, better known as the "alligator salamander." This rather startling name belongs to an animal that is discrete, carnivorous, and has folded skin of darkish color. It still haunts torrents and rivers of the United States such as the Mississippi and the White Rivers in Missouri and Arkansas. Secretive, nocturnal and solitary, *Cryptobranchus alleganiensis* can attain a length of up to 70 centimeters. With its other worldly morphology – half rock, half alligator – the alligator salamander drove early naturalists crazy. "There is an other variety of fish, or whatever one may call it, resembling a small catfish, but having four short legs," wrote David Zeisberger in his 1779–1780 *History of the Northern American Indians* (1910: 74). An old legend claims that this salamander can sabotage fishing lines without being detected by clogging them up with mucus, and that its bite is poisonous. Its totally unmerited reputation earned it the nickname of "Devil Dog." This deceptively calm, occasionally cannibalistic animal lives beneath rocks and sucks in everything that moves: arthropods, worms, snails, or fish. Any animal smaller than itself is potential prey.

The giant American salamander was first described by French naturalist François Daudin in 1803. It gave its name to an entire family, that of the Cryptobranchidae ("those that hide their gills"), which contains the oldest known fossil salamanders (for example, *Chunerpeton*, from the Late Jurassic of China; see fig. 3.32). The American species today has two Asiatic cousins – one in Japan (*Andrias japonicus*), and the other in China (*Andrias davidianus*). *Andrias* is a highly unusual genus: *A. japonicus* (fig. 3.35) is 1.5 meters long and is the largest living amphibian in the world after its Chinese cousin – fatter, heavier, and more imposing at 1.8 meters and weighing in at 25 kilograms! This colossus is equipped with a tail almost as long as its body and its greatly reduced orbits are placed on the upper part of the head, allowing it to perceive changes in luminosity. It detects prey thanks to sensory lines that run the length of its sticky, blotchy skin (a character also found in some ancestral forms). It lunges from a good distance and swallows its victim in the blink of an eye.

These two urodeles remain poorly known. We only know that, like their American cousin, the Asiatic forms live hidden in cool rivers (the Yangtze or Yellow rivers in China, for example) and come out only at night. In August, during the mating season, hundreds of individuals battle upstream, gathering to couple: a female can produce up to 400 eggs, which the males fertilize and jealously protect until they hatch in spring. Once sexual maturity is reached, adults conserve some larval characteristics all their lives (neoteny).

Victims of habitat destruction and much in demand in some unscrupulous Hong Kong restaurants, giant Asiatic salamanders are protected today. But is it already too late? Although they are venerated in Japan and an inspiration for numerous manga and cartoons, populations of this wonderful gluey monster are at critical levels. This is a shame, as this animal of strange habits enthralls scientists who have only just managed to breed them in captivity.

Together with photographers, museographers, and scientists, we have initiated a Franco-Japanese project for the study and reintroduction of specimens of *Andrias japonicus* into French zoological centers. The objective is to stimulate research concerning the history, development, and conservation of this salamander of which there is so much still to learn. *Andrias japonicus* is a panchronic species just like the famous coelacanth (see chapter 1), and is very close to fossil species such as *Chunerpeton tianyiensis* (see fig. 3.32). *Andrias* is, therefore, to lissamphibians what the coelacanth is to the sarcopterygians! In addition to its external morphology, *Andrias* represents an interesting model to better understand the locomotion, behavior, and lifestyles of some stegocephalians, notably those with short, rounded snouts (scientists talk of ecomorphotype). We have everything to gain from protecting this species before it is too late. "The [Native] American who first found Christopher Columbus made an unfortunate discovery," as Georg Christoph Lichtenberg once wrote. The same could be applied to *Andrias*: the first one that found humans made an unfortunate discovery.

3.35. *Andrias japonicus*, "sumo" salamander. At 1.5 meters long, it is almost the largest living amphibian in the world: only its Chinese cousin, the salamander *A. davidianus*, outstrips it (at 1.8 meters and 25 kilograms)! We still know little about the biology of these strange amphibians. It would be interesting to know their lifespan in the wild, bearing in mind that the oldest captive specimen died at the age of 52. Over and above this question, the study of giant salamanders is even more important in that it has morphological and ecological similarities to the stegocephalians with short, rounded snouts. *A. japonicus* and *A. davidianus* are seriously threatened by human activity. It is urgent to protect them!

Photo © Daniel Heuclin/Biosphoto.

The Pangaean Chronicles

BETWEEN EARTH AND SKY 4

WITH STEGOCEPHALIANS (notably the temnospondyls and lepospondyls) and the first lissamphibians, we have focused our efforts on amphibians (in a broad sense of the term): all the animals encountered in chapter 3 were, at least regarding reproduction, inextricably linked to the water. Let us leave the Carboniferous, Permian, and Triassic for a moment and look at the world today. Of the living tetrapods, only 4000 species are lissamphibians whereas there are more than 17,000 species of mammals, birds, crocodiles, turtles, lizards, snakes, and other groups that belong to the amniote clade. These tetrapods, unlike their amphibian cousins, do not need to reproduce in water: their embryos develop in a membranous bag, the amnios, where they are enveloped in amniotic fluid. Amniotes represent more than 80% of living tetrapods, and appeared in the fossil record shortly after the first stegocephalians, 310–315 Ma (Late Carboniferous). This chapter attempts to lay out the first pages of their evolution. Which of the stegocephalians is most closely related? What did the first ones look like? How did those living in forests evolve during the Permian and Triassic? How were they affected by the great mass extinction the Earth experienced between these two geological stages? There are many questions to address, after which we will go on to describe the morphological audacity that characterized tetrapod evolution, long before dinosaurs stole the limelight.

The past is never dead. It's not even past.

William Faulkner (1897–1962)

The Origin of Amniotes

The closest ancestor of the amniotes is to be found among Paleozoic stegocephalians and, more precisely, within the reptiliomorphs. Fewer in number and less diversified than batrachomorph stegocephalians (which, as we saw in chapter 3, are more closely related to lissamphibians than are other stegocephalians), the reptiliomorphs are composed of three principal groups: the anthracosaurs, the seymouriamorphs, and the diadectomorphs.

According to Michael Benton of Bristol University (United Kingdom), the reptiliomorphs are characterized by narrow premaxillary bones (positioned forward in the upper jaw), long and posteriorly pointed vomers (palatal bones), and five digits composed of the following numbers of phalanxes: 2, 3, 4, 5, 4–5. This phalanxial formula is easy to decipher; it simply means that digit I possesses two phalanxes, digit II has three, and so on, until we arrive at digit V, which can have four or five phalanxes. Another character sometimes observed in reptiliomorph stegocephalians is that

The sky seen from the Earth in the Late Permian. *Araeoscelis* is startled by *Coelurosauravus* as it glides past.

4.1. An anthracosaur vertebra (*Archeria crassidisca*, Early Permian, Texas). The vertebrae of these reptiliomorphs are of a particular type known as "embolomere": the pleurocentrum and intercentrum are cylindrical and almost of the same size.

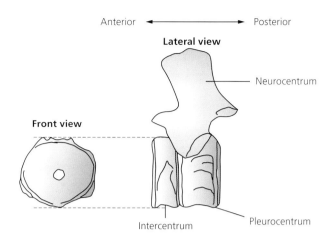

4.2. The strange chroniosuchians. They were reptiliomorphs – that is to say, more closely related to amniotes than to other stegocephalians. (A) *Chroniosaurus dongusensis,* from the Late Permian of Russia, with its elongated snout and ornamented dorsal plate row. (B) Anterior part of the skeleton and skull in lateral right view of *Chroniosuchus paradoxus* (250 Ma). (C) Reconstruction of *Chroniosuchus paradoxus.* Chroniosuchian morphology evokes that of living crocodiles and caimans. The skull of *Chroniosuchus* displays large natural openings between the orbits and nostrils (called the preorbital openings).

Photo reproduced with the kind permission of Jozef Klembara (Comenius University, Bratislava, Slovakia).

the sacrum (pelvic section of the vertebral column) is composed of not one, but two vertebrae, reinforcing the basin – this is useful in terrestrial locomotion. Although all these characters are shared, the reptiliomorphs do not form a homogeneous group from an evolutionary standpoint. However, the reptiliomorphs (stegocephalians + amniotes; see chapter 3) form, according to some authors, a clade.

Which, out of the anthracosaurs, seymouriamorphs, or diadectomorphs, is the sister group of the amniotes? As we will see, this is still open to dispute: the seymouriamorphs and diadectomorphs lead the field, with an increasingly clear advantage in favor of the latter. Discussions among paleontologists on this subject are less animated than those on lissamphibian origins (see chapter 3).

"Coal Lizards": The Anthracosaurs

The term "anthracosaur" (from the Greek "sauros," lizard, and "anthracos," coal) was proposed for the first time in 1934 by Säve-Söderberg, the

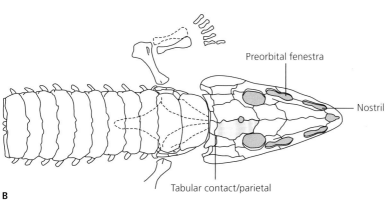

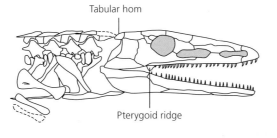

Chroniosuchus

Age Late Permian
Location Russia
Size Pluridecimetric
Features Rather high skull; long snout; lateral orbits; dorsal "carapace"
Classification Anthracosaurian stegocephalian

Swedish paleontologist who discovered *Ichthyostega* and *Acanthostega* (see chapter 1). We gather under this name the stegocephalians—often quite large—from the Late Carboniferous and Early Permian coal deposits of Russia and North America, as well as some forms from the Early Triassic. A new anthracosaur has recently been found in the Permian of Laos, the result of an expedition organized by yours truly and which extends the dispersal of this group even farther eastward. This new fossil, still in the process of description, seems to resemble certain Russian forms, suggesting a connection between Laurasia (the north of Pangaea) and the Indo-Chinese Plate (which corresponds to present-day Laos).

What are the distinguishing traits of anthracosaurian morphology? Their vertebrae are very odd: they are composed of two parts (pleurocentrum and intercentrum), which are cylindrical and frequently of the same size (fig. 4.1). When preserved, the five-digit hindlimbs have the following phalangeal formula: 2, 3, 4, 4, 5.

Anthracosaur skulls present several notable characteristics (fig. 4.2): the tabular bone (in the back skull) is often elongated backward forming "tabular horns" of variable size (a character shared with seymouriamorphs and amniotes) and which contacts the parietal, whereas the supraoccipital (at the base of the skull) is sufficiently developed to be visible in dorsal view. Anthracosaurs have relatively closed skulls: they have neither post-temporal (behind the temple) nor palatal fenestrae (interpterygoidian fenestrae). The latter, if at all present, are reduced to mere slits. The pterygoid, a palate bone, is so ventrally developed that it forms, in this group, a blade that is visible under the skull when viewed laterally (a character shared with seymouriamorphs and amniotes). The stapes (or tympanum, a stick-shaped bone located at the back of the head) is robust and short and considered to be archaic.

Anthracosaur skulls are a veritable mosaic; some characters appear unique to the group and others are shared, either with seymouriamorphs and amniotes or with more "archaic" stegocephalians. It is hardly surprising that the phylogenetic position of the anthracosaurs remains uncertain.

The peak of anthracosaur diversity was reached in the Permian and the group survived the great Permian–Triassic mass extinction (to be treated in greater detail below). How did these stegocephalians make it through this catastrophe? The key to the enigma lies, perhaps, in their lifestyle. Carnivorous and longirostral, their overall morphology is reminiscent of living crocodiles and caiman so, perhaps, they occupied the same ecological niches. Very specific habitats (lakes and confined bodies of water) and the special position crocodiles occupy in the food chain (these predators can survive without eating for several months and often prefer rotten to fresh meat) are often cited as reasons for their surviving the Cretaceous–Tertiary mass extinction 65 million years ago. Perhaps in the same way, thanks to their analogous lifestyle, this is how the anthracosaurs survived the great Permian–Triassic mass extinction.

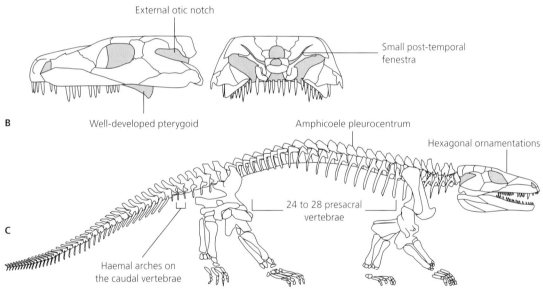

Stegocephalian Cowboys: The Seymouriamorphs

The seymouriamorphs deserve close attention. The characteristic genus of this group, *Seymouria*, was discovered in Texas in Permian layers located near Seymour. This historic town in the northwest of the state was named in 1879 by pioneers from Oregon, in honor of a local cowboy called Seymour Munday.

Morphologically more variable than anthracosaurs, seymouriamorphs are present in the fossil record from the Carboniferous–Permian

4.3. *Seymouria baylorensis*, a Permian seymouriamorph from Texas: (A) cast of a very well preserved skeleton in dorsal view; (B) skull in lateral left and occipital views; (C) complete skeleton.

Photo reproduced with the kind permission of the Muséum National d'Histoire Naturelle/Denis Serrette.

4.4. *Seymouria baylorensis:* this 60-centimeter-long reptiliomorph was capable of lifting its body off the ground, just like its amniote cousins. It could certainly get around on land.

Seymouria

Age Early Permian
Location United States and Germany
Size Up to 60 cm
Features Triangular skull; robust girdles; parasagittal limbs
Classification Seymouriamorph reptiliomorph

boundary until the Latest Permian. These stegocephalians apparently did not survive the hecatomb of the Permian–Triassic. Today we know fewer than 10 genera from North America, Europe, and Asia (corresponding to Laurasia in the Permian).

Like most stegocephalians, Seymouriamorphs (figs. 4.3, 4.4, and 4.5) are characterized by an ornamented skull often showing specific patterns. The trunk is relatively short, comprising 24 to 28 vertebrae that are composed of a cylindrical amphicoelous pleurocentrum (backward-facing concavity). They have haemal arches (small riblike structures) on the first caudal vertebrae, which would have greatly stiffened the tail.

The posttemporal opening in the skull, closed in anthracosaurs, is merely reduced in seymouriamorphs. As in anthracosaurs, the tabular bone contacts the parietal (also an amniote character) and the pterygoid is vertically developed into a blade, visible when the skull is laterally viewed. In addition, seymouriamorphs possess a foramen (a small hole) in the front of the mandible.

As in anthracosaurs, this mosaic of characters makes the precise phylogenetic position of the seymouriamorphs within the early tetrapods a subject of debate, and one that has been raging for nearly a century. Previously considered as sister group of the amniotes, they are now considered the sister group of the diadectomorphs or of the diadectomorph + amniote assemblage. The matter is unresolved, especially given that the limbs of some seymouriamorphs are parasagittal as they are in the majority of diadectomorphs: they were held under the body during walk-

4.5. *Discosauriscus austriacus,* a seymouriamorph from the Carboniferous–Permian of the Czech Republic. This stegocephalian is characterized by its large orbits and a flat, triangular skull. It is known from several hundred larval and juvenile specimens fossilized in lake sediments. Unfortunately no late adult individuals have been found; perhaps they metamorphosed into terrestrial adults and did not fossilize in the same environment. (A) A remarkably well preserved skull in dorsal view; (B) reconstruction of the animal; (C) an almost complete skeleton in dorsal view.

Photos reproduced with the kind permission of Jozef Klembara (Comenius University, Bratislava, Slovakia).

ing (and not splayed laterally). This erect posture is seen in amniotes (such as crocodiles).

Let us now look more closely at two particularly interesting seymouriamorphs, starting with *Seymouria baylorensis* (fig. 4.4), type species of the genus. This modestly sized stegocephalian of monitor lizard–like appearance is known from the Texas Permian. Its skull does not show the dermal sensory canals that characterize an aquatic lifestyle. Its limbs are long and robust, suggesting terrestrial locomotion, and no longer aquatic or amphibious like those of most other stegocephalians. For these reasons *Seymouria* used to be classified as a basal reptile, and therefore an amniote. We now know that it was a non-amniotic tetrapod; therefore, it reproduced in water.

Discosauriscus

Age Late Carboniferous–Early Permian–300 Ma
Location Czech Republic and France
Size Up to 30 cm
Features Large, flat, triangular skull; large orbits; rather short limbs
Classification Seymouriamorph reptiliomorph

Between Earth and Sky   117

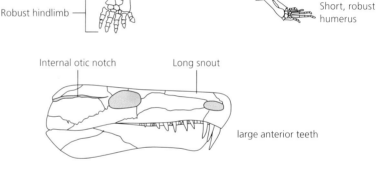

4.6. (A) A fight between a young *Limnoscelis* (in the water) and a young *Diadectes* (thrown into the air), two diadectomorphs from the Early Permian of North America. According to many paleontologists, the sister group of amniotes is to be found within the diadectomorphs. (B) Skeleton and skull of *Limnoscelis* in lateral left view. This 2-meter-long diadectomorph possessed a very triangular skull, relatively closed (the otic notches are internal), as well as robust limbs and girdles, suggesting a terrestrial locomotion. Its snout is somewhat doglike, also resembling that of the "mammal-like reptile" *Cynognathus*.

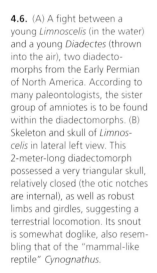

Limnoscelis

Age Early Permian
Location United States
Size Up to 2 m
Features Triangular skull; long snout; large anterior teeth
Classification Diadectomorph reptiliomorph

Discosauriscus is the second seymouriamorph genus we will examine. It is found in the Carboniferous and Permian deposits of Europe. Its external morphology suggests salamander more than small reptile. *Discosauriscus austriacus* (fig. 4.5), discovered in the Late Carboniferous–Early Permian of the Boskovice Basin (Czech Republic), is particularly interesting. There are several hundred extremely well preserved specimens carefully housed at the Comenius University and National Museum of Natural History of Bratislava. This has allowed the morphological characters of *D. austriacus* to be studied in the minutest detail by my Slovakian colleague, Jozef Klembara. Microscopic examination of the skull bones has revealed the existence of subcutaneous electro-receptors. It is one of the first times that such structures appeared in tetrapods. Perhaps it allowed our seymouriamorphs to defend themselves by emitting electrical discharges, or perhaps it was a means of orientation in the water.

Klembara has also identified several growth stages among the numerous fossil specimens. There is a plethora of larval or juvenile

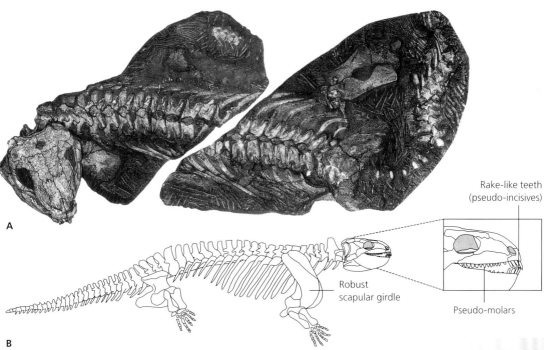

4.7. The first herbivorous tetrapod, *Diadectes* from the Early Permian of North America. This "bovine" reptiliomorph (see fig. 4.6A) possessed rake-like teeth with "molars" behind—useful for browsing great quantities of plants. This dentition is as specialized as that of later herbivorous amniotes such as *Moradisaurus* (see fig. 3.15). (A) An almost complete specimen in dorsal view (some limbs are missing and the cranial sutures have been redrawn on the photograph). (B) The skeleton in lateral right view. The enormous girdles suggest a terrestrial locomotion.

Photo reproduced with the kind permission of Stuart Sumida (California State University, San Bernardino).

Diadectes

Age Early Permian
Location United States
Size Up to 3 m
Features Herbivorous teeth; robust skeleton; short skull
Classification Diadectomorph reptiliomorph

Between Earth and Sky

individuals but not one single adult has yet been identified. How can this be? The fossils that have been studied display small dermosensory canals on the surface of the skull bones, clearly indicating the aquatic lifestyle of larval and juvenile *Discosauriscus*. The Boskovice Basin sediments correspond to those of an ancient lake system. Did those aquatic larvae, once mature, transform into terrestrial adults?

With Sophie Sanchez, we carefully analyzed long bone sections from 12 *Discosauriscus* specimens in the hope of picking up any paleobiological message recorded in their internal, histological structure (see chapter 5). The results were surprising and have provided essential information on the life cycle of a species that is 300 million years old. Importantly, they show that *Discosauriscus austriacus*, with a lifespan of approximately 20 years, attained sexual maturity at around 10 years old, similar to living salamanders. This is even more surprising given that this species is phylogenetically closer to amniotes than lissamphibians.

Browsing Reptiliomorphs: The Diadectomorphs

Diadectomorphs are the reptiliomorphs most often cited as the closest ancestors of the amniotes. Known from the early Late Carboniferous to the late Early Permian uniquely in Euramerica, their origin and diversification are still contentious subjects. They are sometimes considered the sister group of the seymouriamorphs and sometimes to the amniotes with which they share common characters in the back skull region and the cervical complex (atlas and axis). These similarities struck nineteenth-century paleontologists, who, therefore, classified them as reptiles ("cotylosaurians"). Other authors (such as Dave Berman), believe this group should be included in the amniotes, where it would be closely related to the synapsids (including the "mammal-like reptiles").

Anatomically speaking, these intriguing reptiliomorphs are characterized by a single postparietal bone in median position (this bone, situated at the back of the skull, is "doubled" in other stegocephalians), an otic notch (invagination of the skull enclosing the stapes or tympanum) principally located in the opisthotic (a basicranium bone), an intercentrum (vertebral element) equipped with strong forward-facing outgrowths and, finally, a very short and robust humerus (fig. 4.6).

Fossil diadectomorphs are fairly rare (much rarer than temnospondyls, for example), but very diverse. The group comprises large forms (up to around 3 meters) both amphibious and terrestrial. *Seymouria*, for example, was equipped with parasagittal limbs that allowed it to get around on land.

Diadectomorphs also had a wide trophic spectrum: some were carnivorous and others herbivorous, a first in tetrapod evolution and, indeed, the first time that any stegocephalians adopted a strict herbivorous diet (it is more appropriate to use the term "vegetarian" diet, as herbaceous plants had not yet appeared in the Paleozoic). Let us consider the example of *Diadectes* (fig. 4.7). This 3-meter-long diadectomorph, found in the Early Permian of North America, holds a double record: it is one of the first terrestrial stegocephalians to attain such a size and one of the first herbivorous tetrapods. It is characterized by a skull made of thick bones, with relatively large otic notches protecting a well-ossified stapes. As in its cousin *Limnoscelis*, the vertebrae and ribs are robust with short, well-ossified limbs; and the massive girdles point toward terrestrial locomotion. However, the most original thing about this stegocephalian is its highly specialized dentition: its front teeth are spatulate, just like real incisors, forming a sort of comb that could rake in large quantities of vegetal matter. Its back teeth, wider and stouter, look almost like molars. They show micro-wear on the surface, a sign of active occlusion (locking and rubbing between the lower and upper teeth): they would have been excellent plant crushers. To top it off, *Diadectes* also possessed a partial secondary palate, allowing it to calmly chew its food and breathe at the same time: a true browser from the Early Permian!

If it proves correct that diadectomorphs contain the sister group of amniotes, it is unlikely to be the *Diadectes* genus. I would lay odds on a diadectomorph that is carnivorous and therefore less specialized. This is, in any case, what amniote evolution suggests, where candidate sister groups of dinosaurs or mammals most often correspond to carnivorous and generalist forms.

The Dawn of a New Era: The First Amniotes

Around 310 million years ago, in the Late Carboniferous, the luxuriant equatorial forests at the heart of Pangaea, bordering the Hercynian orogenic belt, were the venue of an enormous evolutionary event—the invention of the amniotic egg. This egg was probably delimited by a supple, semi-hermitic membrane or by a shell that protected the embryo while allowing gaseous exchanges. Most importantly, the tiny tetrapod contained within was surrounded by an internal membrane, the amniotic sac, and bathed in a liquid, amniotic fluid (fig. 4.8). Thus, embryo development was freed from the aquatic environment, a considerable physiological

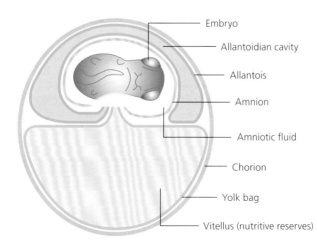

4.8. The amnios: revolution in evolution! The amniote embryo is surrounded by a membrane, the amnios, delimiting a cavity filled with amniotic fluid: it develops in a liquid environment. With this evolutionary innovation, tetrapods are no longer linked to water for reproduction; they can lay their eggs elsewhere! The allantoidian cavity, delimited by the allantois, isolates the embryo from its organic waste. The vitellin or yolk bag contains nutritive reserves (vitellus). The embryo and all its envelopes (known as embryonic annexes) are protected by a soft, semi-hermetic membrane or shell (not shown here) which allows gaseous exchanges.

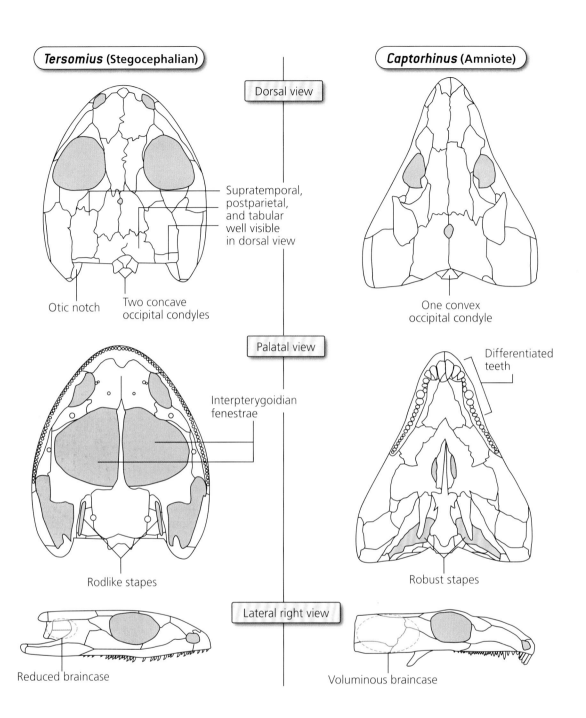

4.9. Cranial comparison between a stegocephalian (*Tersomius*) and an amniote (*Captorhinus*). The amniote lacks otic notches, interpterygoid fenestrae, and has only one occipital condyle. Its braincase is more voluminous than that of the stegocephalian, its teeth are differentiated and it possesses robust stapes.

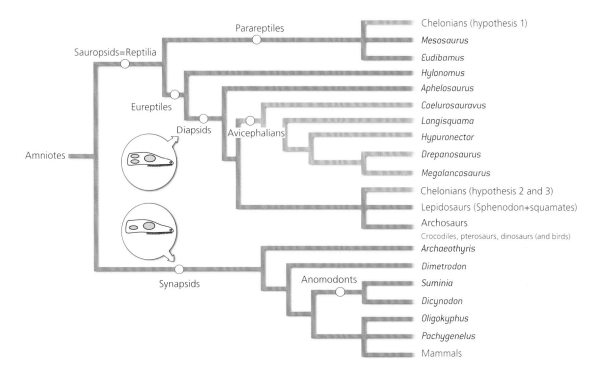

4.10. Evolutionary tree of amniotes. Their classification is mainly founded on the number and position of temporal fenestrae in the back of the skull: diapsids have two pairs of fenestrae, synapsids one. Branches, the positions of which are disputable, are indicated in yellow. Living taxa are in red.

innovation because, for the first time in their history, tetrapods could lay their eggs elsewhere other than in water! The amniotes were born.

Amniotes and Co.

Amniotes emerged in the Late Carboniferous. Nevertheless, no amniote fossil egg has been found in Paleozoic layers. In the twentieth century, the American paleontologist Alfred Sherwood Romer was convinced he had found eggs in Carboniferous sediments, but these proved to be nodules—in other words, ovoid mineral concretions. It is probable that the first amniotic eggs were quite soft (similar, perhaps, to those of snakes) and not liable to fossilize. We must not give up hope, however, that one day the remains of a Paleozoic amniotic egg will turn up in an environment of exceptional preservation. In the meantime, how can paleontologists identify Paleozoic tetrapods as amniotes in the absence of the clade's defining attribute?

Amniotes, of course, possess other specific physiological characters: they generally have lungs with an extremely folded internal structure, thus increasing the surface area available for respiratory gaseous exchanges; their skin is keratinized (covered in varying forms of a fibrous, protein-rich substance called keratin), forming a waterproof barrier with the external environment; and, when breathing, inspiration is facilitated by the intercostal muscles, causing expansion of the torso. All these characters involve soft tissues which very rarely fossilize. This leads us back to the same question: How do paleontologists identify Paleozoic amniotes

Between Earth and Sky

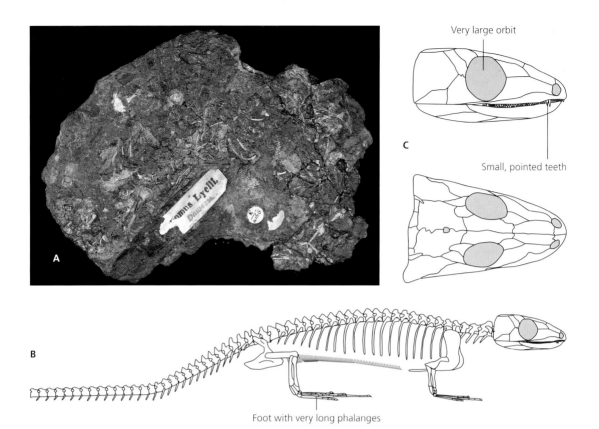

4.11. *Hylonomus lyelli*, the oldest known amniote (sauropsid, early Late Carboniferous – 315 Ma): (A) The historic specimen (disarticulated), discovered in the mid-nineteenth century by Canadian geologist William Dawson at the famous Joggins site (New Scotland, Canada); (B) skeleton in lateral right view; (C) skull in lateral right and dorsal views.

Photo reproduced with the kind permission of Robert Reisz (University of Toronto).

when there is no fossil egg? They rely, quite simply, on other evolutionary innovations that involve the skeleton and do fossilize.

Here is a sample of these "skeletal novelties." The vertebrae of early amniotes closely resemble the "embolomere" type (see fig. 4.1). The neck is often long and highly flexible (with clear articulation surfaces on the first cervical vertebra, the atlas), and is attached to the skull by a single occipital condyle – no longer two, as in the majority of stegocephalians. This condyle is no longer concave but convex (fig. 4.9). The braincase is generally more voluminous than that of stegocephalians; the skull no longer has an otic notch. The stapes are relatively large and clearly transmit sound waves; the occipital surface (back skull) is composed of three reduced bones (postparietal, tabular, and supratemporal) that are no longer visible in dorsal view. As for the rest of the skeleton, the ribs are relatively light and joined to the sternum, the ankle displays a clear articulation surface, the sacrum is composed of one to three vertebrae and the phalanxial formulae are of type 2, 3, 4, 5, 3 for the hand and 2, 3, 4, 5, 4 for the foot.

The First Amniotes: *Hylonomus lyelli* and *Archaeothyris florensis*

The amniotes are divided into two main large clades: the sauropsids, reptiles in the modern sense (including birds); and the synapsids, which

include "mammal-like reptiles" (which do not form a clade) and mammals (fig. 4.10). Sauropsids are characterized notably by paired openings in the palate (suborbital openings), whereas synapsids are characterized by a temporal fossa at the base of the skull (bordered by the squamosal, the postorbital, the jugal, and sometimes the quadratojugal), moderately differentiated teeth, a large and downward facing occiput, as well as a long and striated dorsal extension in front of the pubis.

Sauropsids and synapsids both appeared in the Late Carboniferous. The evolutionary history of these tetrapods is 310 million years old and is still going on today!

The oldest sauropsid was found in the middle of the nineteenth century by the Canadian geologist William Dawson in early Late Carboniferous deposits (Westphalian B–315 Ma) from the famous Joggins locality (New Scotland, Canada). It was named *Hylonomus lyelli* by its discoverer. In Greek *Hylonomus* means "forest dweller": indeed, the remains of this "historic" small tetrapod were found in the fossil trunk of a lycopod tree. Strange! Was this a last refuge when trying to escape a mudslide or simply its natural habitat? We do not know at the moment. The species name of this first sauropsid, *lyelli*, was chosen in honor of British geologist Charles Lyell, a mentor of Dawson's and of other naturalists besides. Charles Darwin himself did not forget in December 1831, before leaving on his long voyage on the *Beagle*, to take the first volume of Lyell's *Principles of Geology*.

What did *Hylonomus lyelli* look like? From head to tail its slender body was no more than 20 centimeters long and looked somewhat "lizard-like" (figs. 4.11 and 4.12). Its small pointed teeth were used for chewing insects and millipedes, which it hunted in the Carboniferous jungle. Its predators were, no doubt, larger tetrapods (notably temnospondyls) and large arthropods, like the famous *Meganeura*, a giant dragonfly with a wingspan of 70 centimeters (see fig. 3.3).

4.12. *Hylonomus lyelli*, 315 Ma. Mascot of the Joggins site, this terrestrial sauropsid, the oldest known amniote, looked like a small modern lizard. In the shadow of the large stegocephalians, these lizard-like forms already swarmed in the Carboniferous forests.

Hylonomus

Age Late Carboniferous–315 Ma
Location Canada
Size Up to 20 cm
Features Slender skeleton; long hands and feet; small ovoid skull; large orbits
Classification Protorothyrid reptile

Between Earth and Sky   125

A

4.13. (A) *Archaeothyris*, the oldest known synapsid (310 Ma). This small terrestrial amniote resembled a monitor lizard. (B) Skull in lateral left and dorsal views (some sutures have not been drawn, not having been preserved on the fossil). Discovered in the Late Carboniferous of Florence in Canada, it "announces" the first great radiation of the synapsids ("mammal-like reptiles").

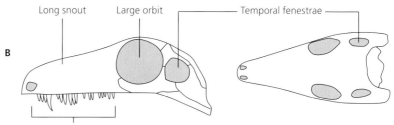

B

Long snout — Large orbit — Temporal fenestrae

Teeth of varied shapes and sizes

Archaeothyris

Age Late Carboniferous–310 Ma
Location Canada
Size Up to 50 cm
Features High triangular skull; large lateral orbits
Classification Ophiacodont synapsid

The title of "oldest amniote" was contested by another fossil, from Scotland this time: *Westlothiana lizzae*–338 Ma. But to the great disappointment of Stan Wood (discoverer of *Balanerpeton woodi*, one of the earliest terrestrial tetrapods [see chapter 3]) and his team, who found it in the Bathgate quarry near Edinburgh, this tetrapod, better known as "Lizzie," was at first considered an amniote but revealed itself to be, after redescription, a stegocephalian. So *Hylonomus* gets to keep its crown, and in 2002 this symbolic fossil was even declared the provincial fossil of Nova Scotia by an Act of Canada's House of Assembly!

What about the synapsids? The oldest of them goes under the bizarre name of *Protoclepsydrops haplous* and was discovered at the same place as *Hylonomus*, the fabulous Joggins site. It is, therefore, also around 315 million years old. Unfortunately *Protoclepsydrops* is so poorly preserved that some specialists question its belonging to the synapsid group. The oldest and most complete synapsid is unarguably *Archaeothyris florensis*, also from New Scotland but from younger deposits than those of the Joggins site: the Florence site (Late Carboniferous, Westphalian D–310 Ma). *Archaeothyris* resembles a small monitor lizard, and is around 50 centimeters long (fig. 4.13). The skull of this carnivorous animal is fenestrated and it has powerful jaws. Its teeth are pointed and vary in size, with two fangs at the front. This last character is a "prelude," to some extent, of the extreme dental differentiation, in both shape and size, which was to appear later within the synapsids.

Separated by several million years, the oldest well-preserved amniotes, *Hylonomus* on the sauropsid side and *Archaeothyris* on the synapsid side, can both be classified in highly morphologically specialized groups: that of the ophiacodonts for *Archaeothyris* ("mammal-like reptiles" known from the Late Carboniferous to the Middle Permian of North America and, perhaps, Europe), and that of protorothyrides (Paleozoic sauropsids) for *Hylonomus*. This classification points to an earlier amniote origin, going back before 315 to 310 million years ago. It would come as no surprise to me if a new amniote were found in Middle Carboniferous or even older deposits in the next few years.

Earth, Wind, and Water: The Great Radiation of the Permian Amniotes

Shortly after its appearance in the Late Carboniferous, the amniote group developed in a spectacular fashion. Due to the rarity of Middle Permian outcrops—this is Olson's Gap—data on their evolution between the Early and Late Permian is scarce. What is clear, however, is that from the Early Permian both synapsids and sauropsids experienced their first evolutionary radiation.

So, on terra firma, ecosystems were dominated by large carnivorous and herbivorous synapsids. At the same time we must not forget that other tetrapods—amphibians, reliant on water for reproduction—start to spread into humid ecosystems: the almost imperceptible lepospondyls and, above all, the temnospondyls, very common in Permian deposits (see chapter 3). As for the amniotes, sauropsids kept a low profile with their modest size, not exceeding several tens of centimeters. They did, however, set out to discover new horizons, aquatic as well as aerial. This is how bipedal, jumping, and even "gliding" forms emerged. With their extremely varied morphologies, some were able to adopt an arboreal lifestyle and so, in the shadow of the great synapsids, invade the upper canopy of the Permian forest.

We are going to tell the fabulous evolutionary story of the Permian amniotes by selecting some interesting examples. Starting with the sauropsids, we will discover forms which could have stepped straight out of the pages of a science fiction novel.

Coming Home: The Mesosaurs, Marine Sauropsids

Most Early Devonian tetrapods were marine (see chapter 1). An amniote colonizing a marine environment could be considered as going back to its roots. The first to attempt this exploit were the famous mesosaurs (or mesosaurids) from the Early Permian of South America and Africa (320–280 Ma), not to be confused with their cousins, the mosasaurs, large varanoid sauropsids which also readapted to a marine environment but appeared later, in the Cretaceous. Incidentally, let's not forget that before mesosaurs, other tetrapods—stegocephalians, for example—had returned to the water, or at least an estuarine environment, from the Late Carboniferous–Early Permian (see chapter 3).

Mesosaurs are of generally modest size (several tens of centimeters on average although one attained nearly 2 meters [figs. 4.14 and 4.15]). The

Mesosaurus

Age Early Permian
Location Africa and South America
Size Up to 50 cm
Features Long skeleton (neck, tail, and snout); large feet and hands
Classification Parareptile

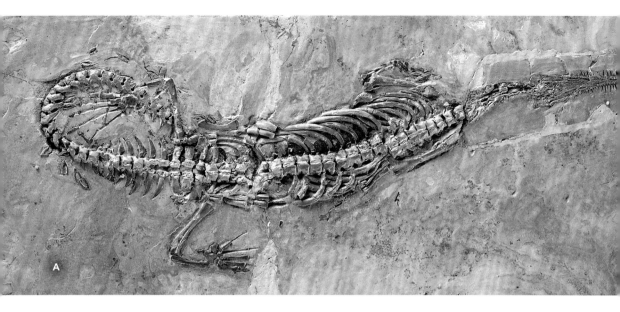

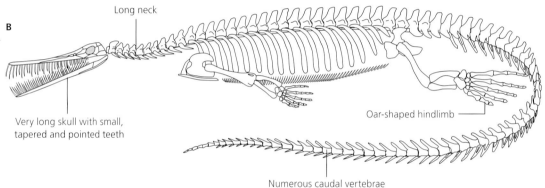

4.14. Mesosaurs, the first amniotes that readapted to water. (A) *Mesosaurus brasiliensis* (Permian): the skeleton in dorsal view is complete but the skull comes from another adult of the same species. The term museographic specimen qualifies this type of composite fossil. (B) Skeleton in lateral left view.

Photo reproduced with the kind permission of Sevket Sen (CNRS/Muséum National d'Histoire Naturelle).

group includes three main genera: *Mesosaurus*, of course, *Brazilosaurus*, and *Stereosternum*. They are already highly adapted to the marine environment. Their very elongated snouts as well as their small, tapered, pointed teeth are useful for capturing small fish or filtering seawater for zooplankton, their long neck composed of 10 or so vertebrae and long tail (for swimming by undulation) and their external nostrils located on the posterior region of the skull are ideally positioned for surface breathing after a long dive. Moreover, some of their thoracic ribs (except in *Brazilosaurus*) show a pachyostosis—that is to say, a strong ossification which increases bone density and, as a result, buoyancy (a character often observed in tetrapods readapted to aquatic life). Their hindlimbs also show long and wide digits, lending them a paddle-like appearance.

Mesosaurus is known today from Permian littoral and marine deposits that extend through Africa and South America (fig. 4.16). If we took away the South Atlantic, these two areas would fit together like a jigsaw puzzle, as would the African and South American coastlines. *Mesosaurus* is one of the key fossil genera that allowed Alfred Wegener to formulate his theory of continental drift at the beginning of the twentieth century:

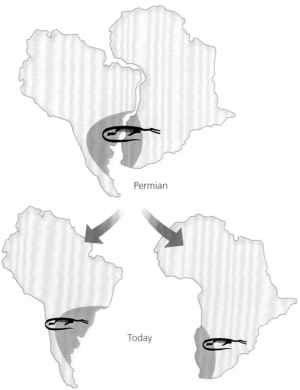

4.15. A mesosaur in the epicontinental sea at the heart of the Permian Pangaea. From a distance they look like young alligators. Not to be confused with mosasaurs, these large Cretaceous sauropsids also readapted to a marine environment.

4.16. African and South American Permian sediments in which *Mesosaurus* is found today (orange zones). Partly based on the symmetry between these two areas of distribution (epicontinental sea in blue) and that of the African and South American coastlines, Alfred Wegener formulated his continental drift theory at the beginning of the twentieth century.

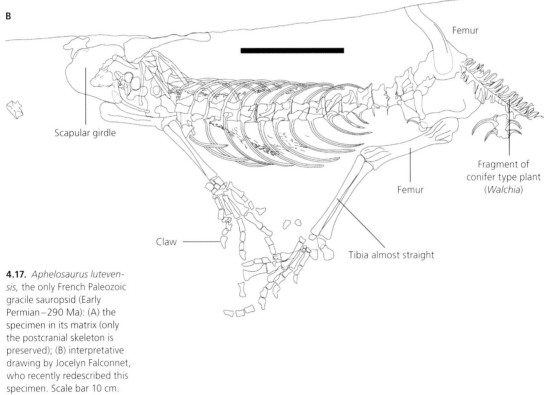

4.17. *Aphelosaurus lutevensis,* the only French Paleozoic gracile sauropsid (Early Permian – 290 Ma): (A) the specimen in its matrix (only the postcranial skeleton is preserved); (B) interpretative drawing by Jocelyn Falconnet, who recently redescribed this specimen. Scale bar 10 cm.

Reproduced with the kind permission of the Muséum National d'Histoire Naturelle/Philippe Loubry; A) and Jocelyn Falconnet (Muséum National d'Histoire Naturelle; B).

4.18. *Aphelosaurus lutevensis:* occasional biped or arboreal? Discovered in the nineteenth century in the Lodève Basin and housed at the Muséum National d'Histoire Naturelle (Paris), this basal diapsid was perhaps an occasional biped. Indeed, it has long limbs, and arms only slightly shorter than its legs. With its large, sharply curved claws, it would also have been capable of climbing trees.

Aphelosaurus

Age Early Permian – 290 Ma
Location France
Size Up to 50 cm
Features Gracile skeleton; rather long limbs terminating in claws
Classification Araeoscelid diapsid

Between Earth and Sky

4.19. *Microsauripus acutipes*, majestic trackways from the Early Permian of the Lodève Basin, France. For the moment, this animal is known only from its footprints!

Photo reproduced with the kind permission of "Uncle" Georges Gand (University of Burgundy, Dijon).

4.20. (facing) Excavations in the Lodève Basin (near Brénas). Several large fossil bones have recently been found at this site reputed for its exquisite fossil traces and trackways. Following this discovery, international excavations were organized to look for other skeletal remains. This systematic search is being carried out thanks to the support of the University of Montpellier, in collaboration with the University of Freiberg (Germany) and with the help of Georges Gand (in the blue shirt, photo B) and Monique Vianey-Liaud. Photos represent the site (A) before and (B) during the excavations; and (C) the author wielding a jackhammer to break up hard argillites from the Lieude Formation.

Photos reproduced with the kind permission of Jorg Schneider (Freiberg University of Mining and Technology, Germany; C); Sébastien Steyer (CNRS/Muséum National d'Histoire Naturelle; A and B).

he proposed that Africa and America were once united within a single continent, Pangaea, before drifting apart during the fragmentation of this continental mass in the Mesozoic. We now know that Africa and South America did not "drift" on the surface of the globe but, rather, were separated during the opening up of the South Atlantic beginning in the Late Jurassic. *Mesosaurus*, therefore, lived in an epicontinental sea within Permian Pangaea, at the center of the future ocean (mesosaur means, in fact, "reptile from the middle").

Get Up! Stand Up! The First Gracile Bipedal Tetrapods

As some amniotes plunged back into the water, others stood up: at the end of the Paleozoic, in at least two groups of sauropsids—the araeoscelids, and the bolosaurs—bipedal forms emerged that were capable of walking,

perhaps for a limited period, on their hindlimbs. These are the first cases of bipedalism in tetrapods; at present we do not know of any bipedal stegocephalians. Like readaptation to the water, bipedalism appeared several times during amniote evolution.

The araeoscelids (araeoscelidians) are small carnivorous sauropsids, tending toward insectivory and mostly terrestrial. Known from the Late Carboniferous to the Early Permian of North America as well as Germany and France, they are double record holders: not content to be one of the first bipedal tetrapods, they are also the oldest known diapsids. Despite their basal position in the amniote tree, they are already highly specialized. Their skeleton is light and gracile with fine limbs and forearms (or forefeet) as long as their arms (or legs) – useful for rapid movements. Some tried using their hindlimbs only. This was perhaps the case for *Aphelosaurus lutevensis*, the only (and quite astonishing) French araeoscelid, housed in the Muséum National d'Histoire Naturelle (figs. 4.17 and 4.18).

Aphelosaurus lutevensis is known only from one specimen found in the mid-nineteenth century in the Tuilières quarry of the Lodève Basin (Luteva in Latin, hence the name), south of the Massif Central region. Although the age of the Tuilières Formation – Late Carboniferous or Early Permian – is still under discussion, it seems that *Aphelosaurus* itself dates from the Early Permian (290 Ma). Described for the first time by the French paleontologist Paul Gervais (1816–1879) in 1858, and then redescribed at the beginning of the twentieth century by Armand Thévenin (1870–1918), this fossil has proven to be a real headache, and for good reason: its skull is not preserved. This made life doubly hard back in the old days when our knowledge of early sauropsids (early reptiles) was very limited.

Recently, Jocelyn Falconnet, in his Ph.D. studies at the Muséum National d'Histoire Naturelle, carried out a detailed osteological redescription of *Aphelosaurus*. Some 150 years after its discovery, he has shown that this sauropsid did indeed display araeoscelidian characters, as had already been supposed, but it was not an araeoscelidian like the others. It possesses a combination of characters unique to itself: a scapulocoracoid (scapular girdle element) anteriorly well developed with a granular

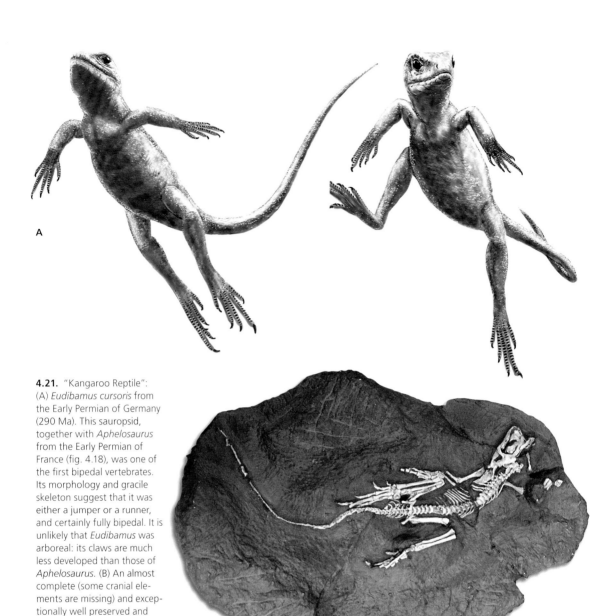

4.21. "Kangaroo Reptile": (A) *Eudibamus cursoris* from the Early Permian of Germany (290 Ma). This sauropsid, together with *Aphelosaurus* from the Early Permian of France (fig. 4.18), was one of the first bipedal vertebrates. Its morphology and gracile skeleton suggest that it was either a jumper or a runner, and certainly fully bipedal. It is unlikely that *Eudibamus* was arboreal: its claws are much less developed than those of *Aphelosaurus*. (B) An almost complete (some cranial elements are missing) and exceptionally well preserved and articulated skeleton. Note the impressively well developed hindlimbs, especially the feet.

Photo reproduced with the kind permission of Dave Berman (Carnegie Museum of Natural History, Pittsburgh).

surface, wrist bones (or carpes) without nervous foramen and of almost identical length, a practically straight tibia, a subtriangular astragal and arms slightly shorter than its legs. All these anatomical oddities call to mind the Jesus lizard (*Basiliscus basiliscus*), a small, living squamata that is gracile and occasionally bipedal. In addition, several arguments converge to suggest that *Aphelosaurus* was perhaps arboreal: its limbs, beneath the body and not on the flanks (parasagittal limbs, capable of articulating in three directions), ended in powerful, sharply curved claws, probably enabling it to climb trees with great dexterity.

Could *Aphelosaurus* cover long distances on its hindlimbs? How much time did it spend in the trees? The paleontologist's job would be made much easier if he could get his hands on more, and more complete, specimens. Unfortunately today the Tuilières quarry is abandoned and it is impossible to get to the layers where *Aphelosaurus* was found. The chances of success are therefore minimal. Nevertheless, visiting the site in the company of Jocelyn Falconnet, we noted the presence of numerous plant fossils of treelike *Walchia*, reinforcing the hypothesis of this small sauropsid's arborical lifestyle.

There is also another part of the Lodève Basin, situated at La Lieude, which could have some surprises in store: in recent decades this site has been reputed for its numerous marks, traces, and imprints of every description—paleoichnes, in scientific terminology—left by Permian tetrapods, including numerous footprints attributed to gracile sauropsids. Of these "light reptiles" known only from their trackways, there is *Microsauripus acutipes* (fig. 4.19). With the help of Georges Gand, international expert on fossil traces, who has made the ichnes of the Lodève Basin his pet subject, we have opened an international excavation site with the aim of finding new skeletal remains from the Permian (fig. 4.20). It is true that *Aphelosaurus* does not come from the same levels as those being scrutinized at La Lieude, but perhaps these excavations—still ongoing—will deliver new gracile sauropsids or other bipeds. Watch this space!

We cannot leave the subject of early bipedal vertebrates without mentioning *Eudibamus cursoris*. Having a short neck, long, tapering legs, and slender arms, this sauropsid was discovered in the Early Permian of Germany (290 Ma), and is remarkable because it was probably 100% bipedal (fig. 4.21). Based on its skeletal characters, paleontologists consider this gracile reptile to belong to the bolosaur group, and it either ran or jumped to get around. *Eudibamus*, unlike its cousin *Aphelosaurus*, would not have been arboreal since its claws were too small and blunt.

An Identified Flying Reptile: *Coelurosauravus*, the Oldest Flying Vertebrate

From the Early Permian, several forms of terrestrial tetrapods experimented with bipedalism. Having done this, they lifted their eyes skyward, because walking on two legs involves raising the vertebral column and, therefore, the head. What animals would *Aphelosaurus* and *Eudibamus* have seen in the sky at that time? Just insects? No, because by the end of the Paleozoic some tetrapods completely abandoned solid ground to explore this new dimension.

Surprising as it may be, the first flying vertebrates were neither birds nor feathered dinosaurs such as *Archaeopteryx* (archosaurs; see fig. 4.10). It was also not the pterosaurs, those majestic flying archosaurs that appeared at the beginning of the Mesozoic (Triassic) and dominated the skies in the Age of Dinosaurs. Well before them, other sauropsids launched

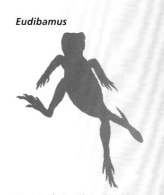

Eudibamus

Age Early Permian–290 Ma
Location Germany
Size Up to 30cm
Features Light skeleton; long legs; large feet
Classification Parareptile

4.22. (A) *Coelurosauravus* from the Late Permian of Europe and Madagascar (250 Ma): the first flying vertebrate. (B) A mostly complete (if rather jumbled) skeleton. With its crest behind the head, this long and flat-bodied amniote (60 centimeters on average) was equipped with long membranous "wings" supported by bony rods in its flanks: this makes it unique in the animal kingdom. These "wings" allowed *Coelurosauravus* to glide long distances well before the appearance of the first pterosaurs.

Photo reproduced with the kind permission of Eberhard "Dino" Frey (Museum of Natural History, Karlsruhe, Germany).

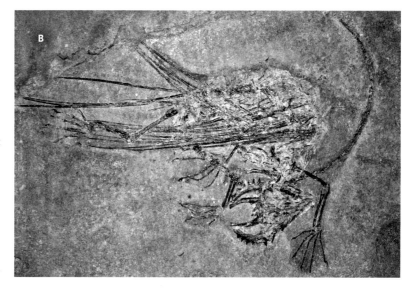

Coelurosauravus

Age Late Permian – 250 Ma
Location Europe and Madagascar
Size Up to 60 cm
Features Very light skeleton (very fine bones); thoracic rods; cranial crest
Classification Avicephalian diapsid

themselves into the blue, experimenting at gliding from the Late Permian onwards. These first flying reptiles have generally been ignored in literature, despite the fact that they allowed amniotes to colonize new higher terrestrial ecological niches, essentially in forested environments. It is fair to say that they have been completely upstaged by pterosaurs, which were the first vertebrates to develop powered flight.

Let us look at the earliest flying vertebrate: *Coelurosauravus* (fig. 4.22). This Late Permian sauropsid, around 60 centimeters long, is known from several complete skeletons belonging to two different species: *C. jaekeli* from Germany and the United Kingdom, and *C. elivensis* from Madagascar. They lived around 250 million years ago – that is to say, 30 million years before the pterosaurs. The genus *Coelurosauravus* is characterized by a morphology unique among tetrapods: its flanks are bordered by long, bony rods of dermal origin on either side of the body; these must have supported two flexible flying membranes. These would have allowed the animal to leap from branch to branch and glide long distances, in the

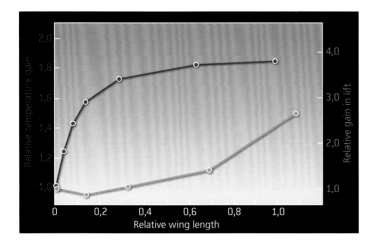

4.23. Wings: adaptation for thermoregulation "diverted" for flight (exaptation). This graph represents the gain in temperature (in red) and in lift (in blue) relative to length of the appendage. The longer the wing the greater the surface area and the more heat it can absorb. After a certain length (between an index of 0.4 and 1.0 here) the temperature gain levels off while lift increases significantly.

From Gould (1985: 94).

4.24. The "Gliders Club." Between hyperextension of interdigital webs (Wallace's flying frog *Rhacophorus nigropalmatus,* and flying gecko *Ptychozoon kuhli*), ribs (flying lizard *Draco volans*) or simply skin (flying lemur *Cynocephalus volans,* flying squirrel *Glaucomys volans,* and synthetic skin of *Homo sapiens* in base jump), living tetrapods display great inventiveness when it comes to gliding. These strategies were already being explored at the end of the Paleozoic and beginning of the Mesozoic.

Between Earth and Sky

4.25. This is not a dinosaur but the best known of the ancient synapsids ("mammal-like reptiles"), *Dimetrodon*: (A) skull in lateral right view; (B) an exquisite skeleton at the Field Museum in Chicago; (C) reconstruction. The name *Dimetrodon* means literally "two sized teeth" because of its cutting teeth in the front and pointed teeth behind. This Permian amniote displays typical pelycosaurian characters like differentiated teeth and temporal fenestrae beneath the orbits, surrounded by the squamosal, postorbital, and jugal. Often confused with a dinosaur, *Dimetrodon* reached up to 3.5 meters in length. Carnivorous, it showed a beautiful sail on its back, supported by long neural spines (like several fossil synapsids). This would have played a social role (recognition, intimidation, etc.) among individuals, but also one of thermoregulation. This latter function leads us to believe that these "mammal-like reptiles" were still ectotherms, whereas mammals are endotherms.

Photo reproduced with the kind permission of the Field Museum, Chicago (John Weinstein, 2005).

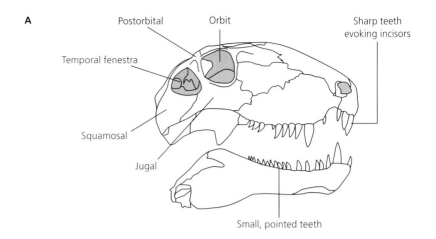

Dimetrodon

Age Permian
Location North America and Europe
Size Up to 3.5 m
Features Robust skull; differentiated teeth; dorsal sail supported by long neural spines
Classification Non-mammalian synapsid ("mammal-like reptile")

same way as a delta plane or a paraglider. *Coelurosauravus*'s "wings" correspond to an entirely original morphological structure: they are neither supported by hyper-elongated ribs, as in the living flying lizard, nor by limb or finger bones, as would be seen later in pterosaurs and chiropters (bats). Instead, its curious bony rods are independent and radially oriented from the thorax between the fore- and hindlimbs.

Were the long membranes of *Coelurosauravus* linked only to flight? Paleontologists have shown that, just like the feathers of dinosaurs, the first insect wings initially played a thermoregulatory role (and are, therefore, exaptations to flight; see chapter 2). We cannot disregard the idea that the initial function of *Coelurosauravus*'s "wings" were to capture heat from the sun's rays, allowing it to warm up more rapidly before going off to hunt insects. We must not forget that we are dealing with a cold-blooded diapsid (ectotherm; the external environment is its principal heat source) and poikilotherm (its body temperature depends on external temperature). It is perhaps only beyond a certain threshold of surface area that the aerial or "gliding" function of these membranes would have overtaken any initial thermoregulatory function (fig. 4.23). In other words, it could also be that these Permian "wings" are an exaptation to flight.

Exaptation or not, *Coelurosauravus* is a unique and hyper-specialized tetrapod. Apart from its "wings," several morphological characters (triangular skull, pointed snout, and enlarged back skull with small crest) have pushed some authors to classify it in a controversial sauropsid group, the avicephalians (literally "bird headed"; see below).

Let us leave our gliding sauropsids for a moment with the observation that many living arboreal species either jump or glide as a matter of routine (fig. 4.24). Among numerous examples are Wallace's flying frog, *Rhacophorus nigropalmatus*; the gliding lizard, *Draco volans*; the flying snake, *Chrysopelea*; the flying gecko, *Ptychozoon kuhli*; the flying lemur, *Cynocephalus volans*; and the flying squirrel, *Glaucomys*. Tetrapods have proven themselves highly inventive in their conquest of the skies, as they navigate from branch to branch and tree to tree.

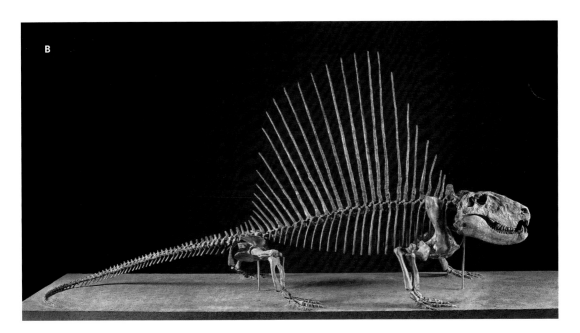

Permian Synapsids: Pelycosaurs and Dicynodonts

In the Early Carboniferous, the first terrestrial vertebrates are quadrupedal. Thanks to the discovery (or redescription) of key fossils (see above), we now know that amniotes that moved about on two legs, bipedal or arboreal, followed hot on their heels (in the Late Carboniferous–Early Permian). Then, in the Late Permian, they experimented with gliding flight. Thus, the near totality of the vegetal cover – including, perhaps, the forest canopies – was already occupied by vertebrates by the Late Permian.

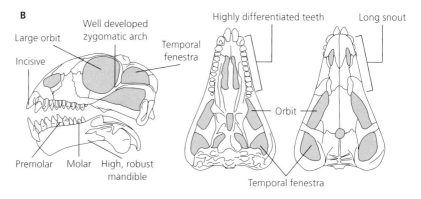

4.26. *Suminia getmanovi*: the first truly tetrapod browser! (A) Beautifully preserved skull completely freed from its matrix (a painstaking job, partly performed under stereo microscope; see chapter 5). Its large orbits and highly differentiated teeth are plain to see. (B) Skull in lateral left, palatal and dorsal views.

Photo reproduced with the kind permission of Robert Reisz (University of Toronto).

Suminia

Age Late Permian – 255 Ma
Location Kotelnich, Russia
Size Metric
Features Highly differentiated herbivorous teeth; triangular skull; large orbits
Classification Non-mammalian synapsid ("mammal-like reptile") anomodont

Was there any reason for this tetrapod exploration of vertical environments? We can entertain two hypotheses: either the swarms of insects living in the trees represented a trophic lure, or fierce competition on the ground favored the natural selection of arboreal forms. It is true that small sauropsids must have been choice prey for giant arthropods as well as for their amniote cousins, the Permian synapsids – where we now focus our attention.

Permian synapsids are "mammal-like reptiles," or non-mammalian synapsids. Although uniquely terrestrial, they occupied various ecological niches. Of these, let us first look at the pelycosaurs, a group comprising mainly large carnivores – 1 to 3 meters long – with a dorsal crest, just like

the well-known *Dimetrodon* (fig. 4.25). These pelycosaurs dominated and terrorized the terrestrial Permian ecosystems while others experimented with an herbivorous diet. This, in itself, is not new within tetrapods because, at the beginning of the Permian, diadectomorph stegocephalians were already oriented toward this diet (see above). The "mammal-like reptiles," however, went farther: Late Permian forms already display a highly specialized system of plant mastication, embodied by *Suminia getmanovi*. With its extremely odd dentition, it is the first truly tetrapod browser: a veritable Permian "sheep" (figs. 4.26 and 4.27)!

Other Permian "sheep" are also well known; the dicynodonts (literally "with two dog teeth"). These therapsids (one of the synapsid clades), widespread and herbivorous, are represented in the fossil record from the Middle Permian to the Late Triassic (perhaps even until the Cretaceous, according to clues recently discovered in Australia). They experienced a radiation at the end of the Permian. Dicynodonts gather around 70 genera, varying in size from that of a rat to that of a full-grown cow. They possess a short trunk and tail, powerful limbs and pectoral girdle, a light but strong skull characterized by greatly enlarged temples (allowing for the insertion of powerful masticatory muscles) and a very short, thick, flat snout. Most were edentate, their narrow jaws terminating with an imposing curved beak—reminiscent of turtles or dinosaurs (ceratopsians or ornithopods, for example)—that sheared plants and then crushed them

4.27. *Suminia getmanovi:* the Permian "sheep." The highly specialized teeth of this modestly sized synapsid from the Late Permian of Russia (Tatarian) allowed it to pulp and slice up its favorite plants before swallowing. Microwear marks on the teeth suggest that the masticatory movements were anteroposterior (front to back).

Between Earth and Sky   141

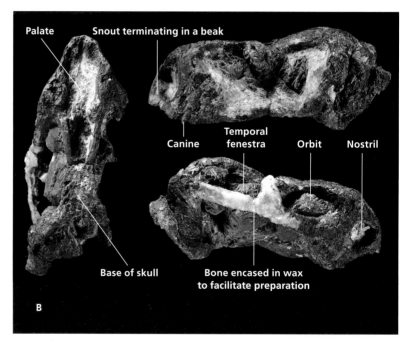

4.28. (A) Three *Dicynodon* (right) surprised by a gorgonopsian (left). *Dicynodon* is an herbivorous synapsid commonly found in continental Permian deposits. (B) Skull from the Late Permian (the preparation of which is still in progress) discovered during a paleontological expedition to the banks of the Mekong River in Laos. Left, palatal view; right, lateral views. The Permian beds of Laos have yielded numerous remains of *Dicynodon* and other terrestrial tetrapods, implying a connection between Eurasia and the Indochinese block from the end of the Permian onward. It is an important point as the layout of this part of Pangaea is still poorly known.

Photos: Sébastien Steyer (CNRS/Muséum National d'Histoire Naturelle).

with vigorous abrasion between the jaws in front-to-back movements. Many dicynodonts also possess two prominent canines (hence the name of the group) that could have been used in courtship display or intimidation of rivals and enemies.

Dicynodon is, as the name suggests, the classic representative. Of modest size (averaging 1.2 m), this genus, known only from the Latest Permian, was very widespread; it has been found in Asia (China and Laos), Russia, and Africa (South Africa and Tanzania, for example; fig. 4.28). It is often considered a good stratigraphical fossil for dating continental Permian beds as it combined maximal geographical distribution with minimal geological duration (or minimal stratigraphical distribution).

At this stage, the end of the Permian (255 Ma), tetrapods were already well established and diversified in various ecosystems; in and under the water, of course, but also on land, where the amniotes even took to occasionally "growing wings"! However, at the transition between the Permian and Triassic (251.4 Ma), major catastrophic events caused dramatic upheavals in paleobiodiversity, a cataclysm of a magnitude unmatched to this day: in the oceans, 96% of species disappeared (57% of families and 83% of genera); on the continents, the devastation was also considerable if harder to estimate – as far as we can judge, 70% of terrestrial vertebrates disappeared. The amniotes (marine and terrestrial reptiles) lost 78% of their families. Geologists and paleontologists call this event the "Permian–Triassic crisis."

"The history of life is not a long, gently flowing river," wrote paleontologist Eric Buffetaut and, indeed, our planet has experienced at least five episodes of mass extinction in the course of geological time. A mass extinction is defined as an event that has led to the disappearance of at least 50% of the total number of species over a period lasting between 10,000 and some million years. Apart from the Permian–Triassic crisis, there has been the Ordovician–Silurian crisis (445 Ma), Frasnian–Famennian crisis (375 Ma), the Triassic–Jurassic crisis (200 Ma) and the well-known Cretaceous–Tertiary (or KT) crisis that, 65 million years ago, saw the extinction of the non-avian dinosaurs (fig. 4.29). Perhaps we should also mention that, according to scientists, we are today on the brink of a sixth major life crisis – this time linked to human activity.

The Permian–Triassic crisis remains the most murderous of them all. It marked the boundary between two eras, two worlds even: the Paleozoic and the Mesozoic. It saw the extinction of large groups of marine organisms like those resembling present-day corals (the graptolites – in

The Mother of Mass Extinctions

Dicynodon

Age Late Permian – 250 Ma
Location Worldwide (Tanzania, South Africa, Russia, China, Laos, etc.)
Size 1.20 m (on average)
Features Fenestrated skull; prominent canines; beak; robust limbs
Classification Cynodont synapsid

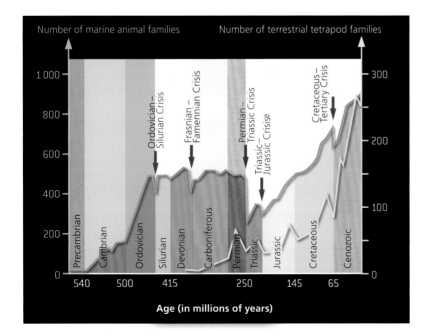

4.29. The global paleobiodiversity curve through the geological timescale, measured in families of marine animals and terrestrial tetrapods. Five major episodes where biodiversity plunged dramatically are recognizable. These correspond to the "Big Five," the five great bio crises our planet has experienced. The Permian–Triassic crisis was the most murderous of them all (after Benton [1999]).

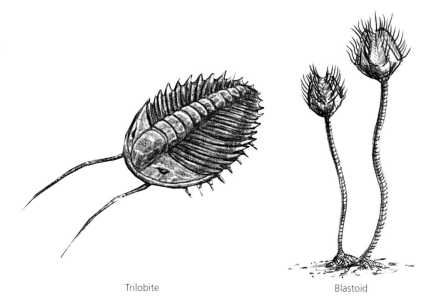

4.30. Two victims of the Permian–Triassic crisis: the trilobites (arthropods) and the blastoids (echinoderms). In total, 95% of marine species died out.

Trilobite   Blastoid

decline since the Devonian, tabulate, and tetracorals) as well as the trilobites (also in decline since the Devonian), the eurypterids ("sea scorpions"), and the acanthodians ("spiny fish"). Other groups flirted with total extinction, including the echinoderms (98% of families disappeared, notably the blastoids), bryozoans (at least 75% of families), brachiopods foraminifers (97% extinction, including that of fusilines) and radiolarians (99% extinction) – both of the latter constituted the plankton of that time. Other groups, such as bivalve and cephalopod mollusks, were partially wiped out. Statistics on marine vertebrate fauna (fish, amphibians, and amniotes) are few and far between, but we know that their biodiversity was altered beyond recognition. The mesosaurs mentioned earlier, for example, did not survive the crisis.

On the continents the roll call of victims was also very long: many plants (notably gymnosperms and seed ferns) and insects disappeared. Amphibians and amniotes also experienced extensive faunal turnover. Entire families of stegocephalians were wiped off the map (as were the lepospondyls and some temnospondyls); others appeared in the Early Triassic (most of the stereospondyls in the temnospondyl group, for instance). Within the synapsids, large herbivores were deeply affected, and pelycosaurs died out. Information about sauropsids is still too scarce for any conclusions to be drawn. It does seem, however, that some lacustrine ecological niches that harbored carnivorous and longirostral tetrapods (some stegocephalians, for example) constituted refuge zones, a little like the freshwater havens that saved the crocodiles' (but not the non-avian dinosaurs') skin during the 65 Ma Cretaceous–Tertiary crisis.

A fungal peak (an increase in mushroom biomass) has been detected just after the crisis, both in continental and marine habitats. This bears witness to the devastation of the event, since mushrooms quickly colonize environments that are rich in decomposing organic matter.

What could have caused such a catastrophe? It was not the work of a lone killer; several culprits joined forces at the end of the Permian to deal life a serious blow. Paleontologists have the tricky task of gathering clues to solve the case. The "usual suspects" are a phase of intense volcanic activity, a general drop in the oceanic level followed by an enormous transgression, an oceanic anoxia, a possible meteorite collision with the Earth, and drastic climate change. This list, like that of its victims, is by no means complete, but we can attempt to apply a few of these potential causes to a global scenario.

The convergence (coalescence) of lithospheric plates, moving around like pieces of a jigsaw puzzle to form Pangaea, caused the closure of the epicontinental seas and a general drop in the oceanic level. This marine regression exposed the continental shelf (extension of the continent beneath the sea surface), killing or drastically affecting the marine organisms living in shallow waters, including the equivalents of our corals and numerous benthic animals.

Sea-level fluctuations would have disrupted and modified ocean currents, thereby changing the global climate. Colossal volcanic eruptions contributed to this climate change. The Siberian Traps and the Emeishan Traps of China, both around 251 million years old, provide clues. Traps are thickly piled layers of volcanic (often basaltic) rocks extending over immense areas: the Siberian Traps covers 200,000 square kilometers (fig. 4.31). At the close of the Paleozoic, megatons of lava spread across the continent (in only 200,000 years, according to estimates). It is not hard to imagine what the effects of giant lava flows, explosions, volcanic bombs, forest fires, and acid rains would have had on the flora and fauna. All this would have been accompanied by the emission of millions of tons

4.31. The deadly Siberian Traps. Effusions of volcanic rocks covered a surface of some 200,000 square kilometers. This event, 251 Ma, is contemporary with the mass extinction that marks the Permian–Triassic boundary. It is probable that intense volcanic activity triggered this life crisis.

Source: University of Bristol (United Kingdom).

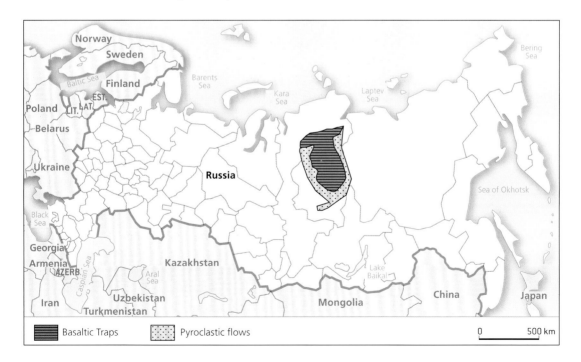

of gases, including greenhouse gases such as carbon dioxide, but also dust and ash hurled up into the stratosphere. These particles would have rapidly spread around the globe, partially blocking out the sun's rays and disrupting photosynthesis. This cloaking of the atmosphere would, at first, cause a drop in global temperatures (this is the "nuclear winter" scenario), followed by intense heating up. This is what happened during the Early Triassic.

To these volcanic events on an unparalleled scale, we can perhaps add a phenomenon of extraterrestrial origin. This hypothesis has already been put forward to explain the Cretaceous–Tertiary crisis, which, in the early 1980s, was thought to have been caused exclusively by the Deccan Traps (in India). Work carried out, in the main, by physicians Walter and Luis Alvarez, however, pointed in another direction: the impact of a massive meteorite could also have been a decisive agent of mass extinction. Their hypothesis was strengthened by the discovery of a massive impact crater, 65 million years old, at Chicxulub on the Yucatán Peninsula (Mexico). It took these researchers more than 15 years to convince the scientific community (notably that of paleontologists) of this scenario, which is now generally accepted. Some scientists wonder if such an event could have also contributed to the Permian–Triassic crisis and have started to look for evidence of a collision. Strangely enough, this has opened up old wounds and arguments over whether this is scientific rigor or jumping on the bandwagon. The upshot is that there are already several candidate craters on the table.

The Bedout Crater, detected in the Indian Ocean off the Australian coast, has caused quite a stir. Larger than Chicxulub, with a diameter of around 200 kilometers, it may have been caused by a bolide the size of Mount Everest colliding with the Earth around 250 million years ago. The impact would have created a shock wave greater than that from the simultaneous detonation of the world's nuclear stockpile. This crater is partly made up of melted rocks that apparently contain minute traces of chromium, an element that, like the iridium associated with the Cretaceous–Tertiary crisis, is extremely rare on Earth. Moreover, rocks that may correspond to explosive residues have been found in Australia, South Africa, and Antarctica. But other geologists consider these to be traces of volcanism, a repeat of the debate that strongly polarized the scientific community about the Chicxulub Crater. It is true that hard evidence of an impact is still lacking, but it should be borne in mind that, compared to Cretaceous deposits, those of the Permian have neither been explored nor studied in the same depth. The Permian–Triassic crisis has yet to relinquish its secrets.

"Natura abhorret a vacuo": Triassic Reconquest

Imagine Pangaea just after the Permian–Triassic catastrophe: a land devastated by forest fires fed by lava flows, acid rains impregnated with ash falls in black droplets to splatter on gray soils covered in a thick pall of mushrooms and lichens. Meanwhile, the climate is heating up and great torrents sweep down the feeble slopes of the vanishing Hercynian Chain. Nevertheless, in the midst of this post-cataclysmic landscape, somewhere deep within lakes and forests, life has resisted destruction. For the survivors (the amniotes for the most part), it is their turn to enjoy evolutionary radiations.

Whatever the period, realm (continental or oceanic), or habitat (pelagic, coastal, or lacustrine), crises have always profited some—and the "profiteers" have not always been those that were the most visible before the catastrophe. In fact, mass extinctions liberate ecological niches that benefit other species. It is important to remember that the opportunists often existed before the crisis, but occupied only a humble place in the ecosystem. It must be added that if ecological niches are liberated, it is not compulsorily followed by evolutionary radiation: this often takes place beforehand. This is a relatively new idea confirmed by data on the Permian–Triassic and Cretaceous–Tertiary crises. In the case of the Permian–Triassic crisis, sauropsids that had kept a low profile during the Paleozoic, due to strong competition from large synapsids, gained the upper hand at the beginning of the Mesozoic. Concerning the Cretaceous–Tertiary episode, paleontological finds show that ancestors of the major placental mammal groups were present before the crisis but remained discreet: ecologically speaking, the dinosaurs and other archosaurs relegated them to the sidelines.

Let us return to the Triassic and ask ourselves what the amniote survivors from the great extinction looked like. We will not examine every group but, instead, we will shed light on poorly known fossil forms that, like avicephalians, had unusual morphologies. We will also look at some that, via the mammals (a possible sister group), have a radiant future ahead of them. We will briefly touch on the more "classic" fossil amniotes—turtles, lepidosaurs, and archosaurs—at the end of the chapter.

4.32. *Otopteryx volitans,* an imaginary taxon of rhinogrades. This animal flies backwards thanks to the flapping of its ears. Its nose is highly developed (a characteristic of the group) and serves as landing gear. The morphologies of very real avicephalians (see figs. 4.22 and 4.34) can hold their own with our imaginary rhinogrades dreamed up by zoologists Gerolf Steiner and Pierre-Paul Grassé; the latter urges caution: "Biologist, my good friend, remember that the most ably described things are not always the truest."

Acrobatic Sauropsids: Avicephalians

Longisquama

Age Late Triassic
Location Kyrgyzstan
Size Decimetric
Features Gracile skeleton; possible hyper-elongated "scales"
Classification Avicephalian diapsid

The forests of the Early Triassic provided oases of life for the amniotes that survived the great extinction. Among the escapees were the "avicephalians," already a timid presence in the Permian and perhaps the most original of amniotes. They form, according to some authors, a small clade of fossil diapsids that includes some very surprising forms that seem to push the limits of "morphological possibility." Their skeleton is very gracile, with a light "aviform" skull (evoking that of a bird)–and some even had toothless beaks! These astonishing similarities led some paleontologists to propose, several years ago, close phylogenetic links between avicephalians and birds. We know today that this hypothesis is false; birds stem from the archosaurs; small, gracile, feathered theropod dinosaurs, to be precise. The characters that birds and avicephalians share are only the result of evolutionary convergences, but the phylogenetic position of the avicephalians, such highly specialized animals, is still debatable: the validity of the clade has also been questioned, since each species of the group could also be placed within different and distinct sauropsid groups.

All this does not detract from these fascinating animals, which, like our fanciful rhinogrades (fig. 4.32), resemble the fruits of a vivid imagination. Along with their birdlike heads, they possessed a prehensile tail

4.33. *Longisquama insignis*, avicephalian from the Late Triassic of Kyrgyzstan: skeleton on a slab displaying very elongated "scales" (more than 10 centimeters in length). These are the subject of hot debates: Are they pseudo-scales or pseudo-feathers? Could they simply be fern leaves deposited on the ground before fossilization?

Photo reproduced with the kind permission of the Paleontological Institute, Russian Academy of Science.

evoking that of a chameleon (which belongs to another clade of diapsids: the squamates) or even a monkey. Some strange forms, such as *Longisquama insignis* (Late Triassic of Central Asia), could possess extremely elongated "scales"—the nature and function of which remain unknown. According to some hypotheses, they could have acted as pseudo-feathers and formed "wings" (figs. 4.33 and 4.34)! Surprisingly, this group, made up of under a dozen Permian and Triassic genera, is often ignored in scientific literature. We shall repair this injustice and review some remarkable Triassic avicephalians.

Our first specimen is the eye-catching *Megalancosaurus*, a small avicephalian of some 25 centimeters discovered in the Late Triassic of northern Italy. *Megalancosaurus*'s most striking character is its long tail, which appears to have been mobile and prehensile (fig. 4.35). It evokes that of some monkeys, except that it terminates in fused vertebrae and a sort of "dart" or claw, a first in tetrapod evolution. This is maybe why the paleontologist Phil Senter classifies *Megalancosaurus* and other avicephalians in a subgroup apart, that of the simiosaurs (literally, monkey-lizards). But that is not all. Its fore and hindlimbs possess two and one fused and opposable digit(s), respectively. With its prehensile tail, it has all the allure of a chameleon. In addition, some bony vertebral extensions (neural

4.34. Reconstitutions of *Longisquama insignis* reflecting three hypothetical "functions" for its mysterious "scales." Could these structures have allowed it to fly (top left)? Perhaps to intimidate enemies or for nuptial parade (bottom left)? Or maybe just leaves deposited around the skeleton before fossilization (right)? The question remains open.

Between Earth and Sky

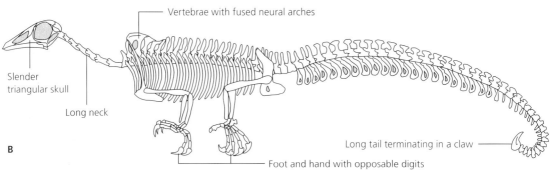

4.35. (A) *Megalancosaurus preonensis*, arboreal avicephalian from the Late Triassic of Italy. Bird-headed, it had a chameleon's feet and a monkey's tail that ended in a claw. This is not a hoax. With its four limbs and prehensile tail, *Megalancosaurus* achieved a degree of specialization equivalent to that of chameleons or primates, but 130 million years earlier! This simian morphology pushed the paleontologist Phil Senter to classify *Megalancosaurus* and other avicephalians in a subgroup called simiosaurs, literally "monkey-lizards." (B) Skeleton in lateral left view (after Matt Celeskey, www.hmnh.org).

Megalancosaurus

Age Late Triassic–210 Ma
Location Italy
Size Up to 25 cm
Features Gracile skeleton; long neck; long tail terminating in a "claw"
Classification Simiosaurian avicephalian diapsid

apophyses) are fused together at the shoulder level, probably allowing the insertion of powerful neck muscles, useful for possible rapid head movements when hunting insects or other small, fast-moving prey.

With its "composite" and unique morphology, it is difficult to imagine the lifestyle of *Megalancosaurus*. Most reconstitutions do seem to agree on its being an arboreal form, perhaps living like a chameleon. Other authors, however, envisage it is as amphibious or terrestrial, hunting on the ground and balancing on its hindlimbs and tail. Some Italian specimens of *Megalancosaurus* do not have the opposable thumb; it could have been an exclusively male character, perhaps for holding the female during mating – or why not an exclusively female character?

Drepanosaurus is another strange avicephalian simiosaur. Like *Megalancosaurus*, it comes from the Late Triassic of Italy. In addition to

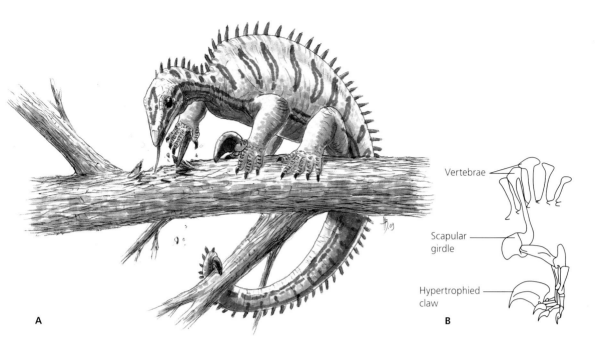

the oddities it shares with its cousin, this diapsid 40 centimeters in length (excluding neck and tail) possessed an enormous, flat, billhook-shaped claw on its forefinger (fig. 4.36). It was the first time in tetrapod evolution – but not the last – that hypertrophy of one or several claws appeared.

Drepanosaurus apparently had the claw only on its "hands." This is often observed in predating, hunting (think of the raptors' claws in *Jurassic Park*), climbing, or burrowing tetrapods. Of these different lifestyles, which did *Drepanosaurus* adopt? Let us look more closely at the morphology of its claw; it is so enormous that it looks disproportionate to the rest of the limb. Just like a bird's beak, the shape of a claw often betrays the use its owner puts it to and the habitat the owner occupies (fig. 4.37). The grizzly (and also the piscivorous dinosaur *Suchomimus*) possesses large, sickle-shaped claws, useful for capturing fish. The mole *Talpa europaea*'s claws are straight and spade like, allowing it to dig tunnels. This does not help us much, in fact, as *Drepanosaurus*'s claw resembles that of neither a bear nor a mole. In fact, it looks like a pygmy anteater's claw. A fairly obscure comparison, I think you will agree. This mammal, also called a myrmidon (*Cyclopes didactylus*) is very shy, nocturnal, and arboreal – with a prehensile tail and a claw like a billhook, which it uses to dislodge insects under tree bark. This is the kind of lifestyle that would have suited *Drepanosaurus* perfectly.

Longisquama was perhaps a "winged" avicephalian, *Megalancosaurus* had a "monkey tail" ending in a "claw," and *Drepanosaurus* had its billhook-shaped claw. Let us turn to North America and New Jersey, where Late Triassic outcrops have yielded another avicephalian, *Hypuronector*, which was equipped with an oar (fig. 4.38). *Hypuronector* is represented by a dozen small fossil specimens (only 10 centimeters long) which are, unfortunately, not as well preserved as their Italian cousins:

4.36. (A) *Drepanosaurus*, an avicephalian from the Late Triassic of Italy. This sauropsid possessed a morphology similar to its cousin *Megalancosaurus* (see fig. 4.35) with, in addition, a claw in the shape of a billhook. This apparent "tool" may have been useful for scratching away tree bark to get at insects, but the function of this hyper-enlarged phalanx remains controversial. (B) Skeleton of front limb and pectoral region (after Jamie A. Headden).

Drepanosaurus

Age Late Triassic – 210 Ma
Location Italy
Size Up to 50cm
Features Large billhook-shaped claw; tail terminating in a "claw"
Classification Simiosaurian avicephalian diapsid

4.37. The "Claw Club." Hypertrophy of the anterior last phalanges—whether knife-, billhook-, or sickle-shaped—appeared several times during tetrapod evolution. Each tool or "hand weapon" is often associated with a specific diet or locomotion: (A) the theropod dinosaur *Suchomimus* fished with its sickles, (B) as does the grizzly *Ursus arctos*; (C) the mole, *Talpa*, digs with its spades, and (D) the pygmy anteater *Cyclopes* hunts insects with its billhooks. The avicephalian *Drepanosaurus*'s billhook (Triassic from Italy; see also fig. 4.36) resembles that of the pygmy anteater.

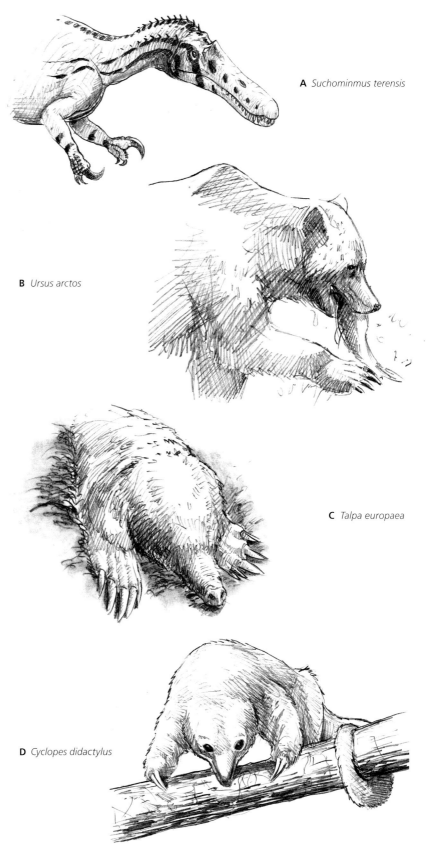

A *Suchominmus terensis*

B *Ursus arctos*

C *Talpa europaea*

D *Cyclopes didactylus*

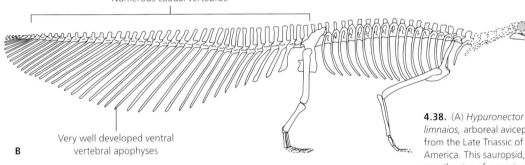

Numerous caudal vertebrae

Very well developed ventral vertebral apophyses

B

4.38. (A) *Hypuronector limnaios*, arboreal avicephalian from the Late Triassic of North America. This sauropsid, which was the size of a praying mantis, was equipped with a tail shaped like a cycad leaf (arborescent fern): Could this be the first example of mimicry in tetrapod history? (B) Reconstructed skeleton in lateral right view. The dotted parts are not preserved but drawn based on known skeletons of similar species (after Matt Celeskey, www.hmnh.org).

at present we do not have the head, neck, or all the digits. However, its tail is well preserved and, surprisingly, it does not terminate in a dart or a claw but in long, fine ventral vertebral apophyses, giving it an oar- or leaf-like appearance. This very slender, tall, flat appendage was larger than the animal's body.

What was its function, if any? Did it act as an oar for swimming? If this were the case, *Hypuronector* would be one of the rare aquatic avicephalians. This is unlikely; the tail, given its flatness and height compared to body size, would have caused too much drag in water – it would have required supersized muscles to operate such a rudder! This appendage does, however, bear a striking resemblance in size and shape and is elongated and asymmetric like the leaf an arborescent fern contemporaneous with the fossil (a cycad close to the living *Cycas*). Perhaps this represents the first-known case of mimicry in terrestrial tetrapods: it is easy to imagine *Hypuronector* moving jerkily along a branch in dappled sunlight, perfectly camouflaged among the leaves, in search of small insects. If this hypothesis proves correct, then *Hypuronector* was an arboreal avicephalian.

Hypuronector

Age Late Triassic–222 Ma
Location New Jersey
Size Up to 12 cm
Features Gracile skeleton; long tail with fine ventral apophyses
Classification Simiosaurian avicephalian diapsid

Between Earth and Sky

4.39. Late Triassic forests were already well structured with diverse ecological communities of non-avian reptiles. From top to bottom: (1) rhamphorhyncoid pterosaurs (at canopy level); (2) *Longisquama* (avicephalian, subcanopy, assuming this form flew); (3) *Megalancosaurus* and (4) *Hypuronector* (simiosauran avicephalians); (5) *Drepanosaurus* (relatively terrestrial avicephalian). These amniotes occupied different ecological niches which varied in height above the ground and food supply (mainly insects).

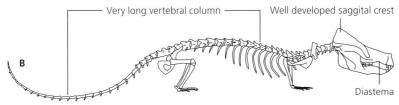
Very long vertebral column — Well developed saggital crest
B
Diastema

4.40. (A) *Oligokyphus,* one of the best-known tritylodonts. Well represented in the Late Triassic–Early Jurassic of Europe, North America, and China, the tritylodonts were small herbivores, measuring between 15 cm and 1 m in length. Together with the trithelodonts, they are among the closest relatives of mammals. (B) *Oligokyphus* skeleton in lateral right view.

The climbing, jumping, and gliding avicephalians we have encountered constituted a veritable arboreal fauna; this was true in the Permian and even more so in the Triassic. These small animals fed mainly on insects. Their fossils can be found in North America (*Hypuronector,* for example) and Asia (*Longisquama,* in Kyrgyzstan), between which there existed, at the heart of Pangaea, a band of forests. Certain zones of this forest could have escaped destruction during the Permian–Triassic catastrophe. They were not as deep or dense as those of the Carboniferous, but they were no less rich for it. In the Permian and Triassic they already harbored numerous small, specialized amniotes that "prefigured" the feathered, scaly, or furry inhabitants of today's forests (such as chameleons, birds, primates, and flying squirrels). Arboreal vertebrate diversity during the Permian and Triassic is still probably underestimated, because forest dwellers in their variety of ecological niches rarely get fossilized, the forest habitat not being as conducive to fossilization as aquatic habitats (lakes, seas, and oceans); and also because many paleontologists prefer to work on dinosaurs! It is, therefore, quite probable that, from the end of the Paleozoic to the beginning of the Mesozoic, vertical forest ecosystems were occupied and structured by different reptile communities (sauropsids), each one specializing in its own habitat height and food supply (notably insects; fig. 4.39). There are many worlds left to discover in paleontology: the Permian and Triassic forests, no doubt, count among them.

Oligokyphus

Age Late Triassic–Early Jurassic
Location Worldwide (United States, United Kingdom, Germany, China, Antarctica, etc.)
Size Up to 45 cm
Features Differentiated herbivorous teeth; slender skeleton
Classification Tritylodont synapsid, close to mammals

A

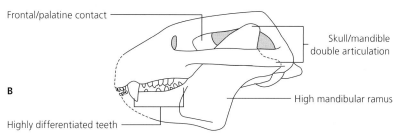

4.41. (A) The trithelodont *Pachygenelus,* from the Late Triassic–Early Jurassic of southern Africa. Together with tritylodonts, trithelodonts are among the closest relatives of mammals. They were insectivores or generalist carnivores. (B) Skull in lateral left view. Two regions of the back skull seem to articulate with the mandible: in the upper region, the dentary touches the squamosal and below, the quadrate contacts with the articular.

Pachygenelus

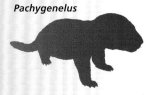

Age Late Triassic–Early Jurassic
Location South Africa and Lesotho
Size Up to 20 cm
Features Ridged postcanines composed of prismatic enamel
Classification Trithelodont synapsid close to mammals

In the Bosom of Evolution: Tritylodonts and Trithelodonts (Synapsids)

Synapsids also traversed the Permian–Triassic crisis but went through a strong faunal turnover. Herbivores appear to have been less affected. Our famous dicynodonts made it through more or less in one piece. But as we have already seen, large carnivorous synapsids – the pelycosaurs – disappeared. They are replaced during the Triassic by large archosauromorphs that already resembled certain dinosaurs (see below). Other smaller synapsids endured in the shadow of the dinosaurs until the end of the Cretaceous and the dawn of the mammals. Among them, two families are candidates for closest mammalian ancestor: tritylodonts and trithelodonts. One of these groups could represent the sister group of mammals, the very bosom from which mammals evolved! Representatives of this group often display a morphology that evokes that of living mammals; their skull is relatively light (the temporal fenestra adjoins the orbit) and simple (prefrontal and postorbital bones are absent). It is also possible that these tetrapods – which were small and, therefore, suffered from increased heat loss – maintained a constant body temperature thanks to a furry skin, another characteristic which brings them closer to mammals. This possible furry covering, together with enlarged orbits, suggests that trithelodonts and tritylodonts were burrowers or nocturnal.

Tritylodonts were small herbivores between 15 centimeters and 1 meter long, excluding tail (fig. 4.40). They are known from the Late Triassic–Early Cretaceous (Aptian), becoming widespread from the Early Jurassic. Their orbital walls resemble those of early mammals. Their dentition is distinctive, with postcanines (cheek teeth) that are very odd, ridged for mashing food; enlarged lower incisors (a bit like a rodent's), and no canines but a diastem (a space between teeth), and rather caniniform incisors. The skull displays a long sagittal crest and a large zygomatic arch, allowing the attachment of powerful chewing muscles, and their highly elongated postcranial skeleton is similar to that of modern mammals.

Trithelodonts were no grazers, but would have probably feasted on insects or were generalist carnivores. They were lower profile than their tritylodont cousins, both in size and spatiotemporal distribution; we only find them in the Late Triassic–Early Jurassic of southern Africa (South Africa and Lesotho) and South America (Argentina). Their morphology is characterized by large postcanines with a well-developed ridge and made of prismatic enamel, as is the case in mammals. These teeth perhaps had a double root, another characteristic that would make them closer to the mammals. In addition, trithelodonts may have a double articulation between the skull and mandible: one between the dentary and squamosal, and the other between the quadrate and articular (fig. 4.41) – the articulation of the mandible in mammals is simple, between the dentary and squamosal. Finally, their small skull (between 3 and 8 centimeters in length) has a frontal bone joined to the palatine, as in mammals. The trithelodont family is much smaller than that of the tritylodonts and, therefore, lesser known (only a few genera have been discovered).

Happy Ending?

Amniotes, like amphibians, survived the Permian–Triassic crisis, but were scarred by the experience: this can be seen in the strong faunal turnover. The large carnivorous synapsids of the Late Permian (the powerful pelycosaurs) disappeared. The top of the food chain was rapidly occupied by other amniotes, sauropsids this time: these were the archosauromorphs (archosaurs and related forms; see fig. 4.10). Behind this name hides a host of very special diapsids, whose evolutionary history goes back to the Late Permian. The only living representatives of archosauromorphs are birds and crocodiles (the other main diapsids, such as the lepidosauromorphs, were also present from the Late Permian or Early Triassic; they are today represented by the squamates – amphisbaenas, lizards, and snakes – and *Sphenodon*).

Looking mostly like large lizards and keeping a low profile in the Late Permian, the archosauromorphs truly thrived after the great Permian–Triassic crisis. In the Late Triassic the group was in full expansion: in the skies, early pterosaurs took their maiden flights; on the ground, the first dinosaurs, initially small and bipedal, skittered timidly around armored herbivores (the aetosaurs) – taking care to avoid phytosaurs, freshwater predators, or ferocious rauisuchians (fig. 4.42). Remember that

4.42. Three reconstituted paleoecosystems of the Late Triassic, based on excavations carried out in (A) the United Kingdom, (B) Morocco (Argana), and (C) Germany. They bear witness to archosaurian diversity at that time. (A) The small *Scleromochlus* (foreground, perhaps at the origin of pterosaurs) fleeing from *Ticinosuchus* in the background (rauisuchian, a large four-legged predator). (B) On the left, *Arganasuchus*, a fierce rauisuchian, threatening *Moghreberia* (dicynodont and the only "mammal-like reptile" represented here) and *Aetosaurus* (aetosaur, a small armored herbivore in the foreground) on the right; in the water, *Palaeorhinus* (phytosaur, crocodile-like piscivore) looks on. (C) Left, the dinosaur *Plateosaurus* (herbivorous prosauropod dinosaur) eating beneath the gaze of two *Procompsognathus* (gracile theropods, carnivorous or insectivorous dinosaurs) on the right.

pterosaurs, aetosaurs, phytosaurs, and rauisuchians were not dinosaurs. They did display, however, striking evolutionary convergences with some Jurassic and Cretaceous dinosaurs (aetosaurs already evoke ankylosaurs, phytosaurs resemble spinosaurs, and rauisuchians look like theropods). They could well have occupied similar ecological niches.

Numerous archosauromorphs were decimated by the great life crisis that marked the Triassic–Jurassic boundary: many of them were wiped out; pterosaurs and dinosaurs counted among the survivors. Dinosaurs would, in their turn, occupy center stage in the planet's ecosystems. But that is another story.

A BRIEF GUIDE TO PALEONTOLOGY

5

HOW DOES A PALEONTOLOGIST RECONSTRUCT the complete animal from an often fragmentary fossil? Contrary to what the misinformed might think, reconstruction is not a product of the imagination but the result of a series of logical steps involving various methods. Some of them date from the beginning of the nineteenth century; others employ state-of-the-art techniques such as microtomography (bombarding a fossil with X-rays so as to "scan" it without damaging it), isotopic analysis, biomechanics, and bone paleohistology. Here is a brief look into the toolbox the modern paleontologist uses to unlock the secrets of treasures found in the field.

> In natural history there are two major pitfalls to avoid: the first is having no system, and the second is trying to attribute everything to one system.
>
> **Georges Louis Leclerc, Comte de Buffon (1707–1788)**

Reconstruct to Better Understand

The first method of reconstruction from fossil fragments is also the oldest. This is comparative anatomy, established by the French naturalist Georges Cuvier (1769–1812). A discipline in its own right, comparative anatomy allows us to reconstruct a complete organism from a single organ or a skeleton from a single bone. It was Cuvier himself who outlined the principle in his *Recherches sur les ossements fossiles de quadrupèdes*:

> Every organized being makes up a whole, unique and closed system, of which the parts fit together and the action of one part engenders a reaction in another. No one part can change without the others also changing; consequently, each part taken separately suggests the nature of the others. Thus . . . if the intestines of an animal are organized for the digestion of flesh . . . its jaws must be made for the devouring of prey; its claws to seize and tear it; its teeth to cut it into chunks; every organ of locomotion built to hunt it down; its sensory organs to detect it from afar. (1812/1992: 97–98)

The comparative anatomy approach is still used today. For example, a tooth found isolated is compared, in size and shape, with the teeth of all animals, both living and from the fossil record. This facilitates a better understanding of the morphology of the animal it belonged to. The nature of the soft tissues—such as muscles and viscera, which rarely fossilize—is deduced directly from clues gathered from the skeleton. For example, nerve or blood pathways are reconstituted from the size, shape and position of foramens, or grooves visible on the bone surface. When the skull is sufficiently well preserved, even the brain can be reconstructed (fig. 5.1)! Indeed, it is possible to deduce its shape and pattern (or proportions of some areas) from analysis of the endocranial cavity

The author extracting a mandible of Moradisaurus *(see pp. 82–84) in the Permian outcrops of Niger in the Sahara.*

Photo reproduced with the kind permission of Christian Sidor, University of Washington.

A

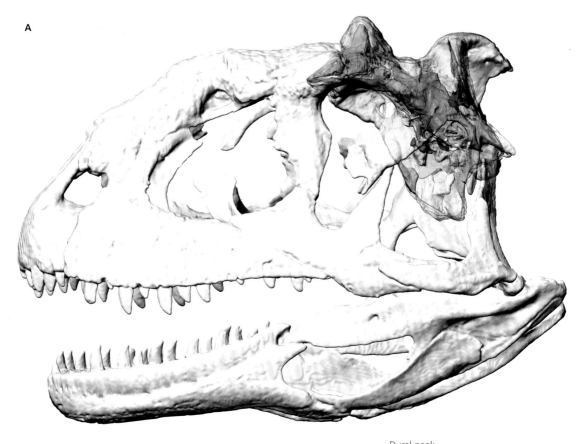

B

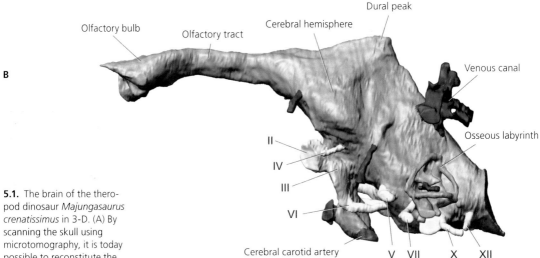

5.1. The brain of the theropod dinosaur *Majungasaurus crenatissimus* in 3-D. (A) By scanning the skull using microtomography, it is today possible to reconstitute the brain of this dinosaur from the Late Cretaceous of Madagascar in 3-D. (in [B] the cranial nerves are labeled with Roman numerals). (C) Reconstruction of the animal's head.

(A and B) Reproduced with the kind permission of Lawrence M. Witmer, Ohio University.

if it has not been too deformed during fossilization. Modern imaging methods, such as microtomography, are essential because they allow us to visualize the interior of a fossil without destroying it. Today these methods are so frequently applied to key fossils of dinosaurs, pterosaurs, and other vertebrates that we can easily employ the term paleoneurology! Muscles are reconstructed by studying the shape and size of the apophyses – that is to say, bone expansions where tendons are inserted. Step by

C

step, comparison by comparison, the initial fossil, often fragmentary, is built up as much as possible and, when there are enough clues, the vital organs are added and it may become a veritable facsimile. To bring a fossil back to life, a paleontologist does the reverse of a coroner practicing an autopsy: the former reconstructs a body as entirely as possible from fragmentary remains, whereas the other "deconstructs" a complete body to extract and analyze its organs.

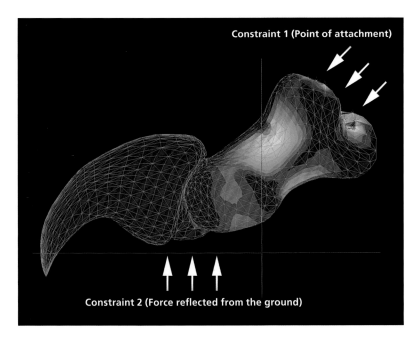

5.2. Modeling the articulation constraints for the clawed thumb of the dinosaur *Plateosaurus* (Triassic, France) while walking. In this 3-D image obtained by finite element analysis (FEA), we can see that mechanical constraints are unevenly distributed along the bone surfaces. They vary according to the forces exerted during walking (arrowed).

Reproduced with the kind permission of the Muséum National d'Histoire Naturelle/ Florent Goussard.

A Brief Guide to Paleontology

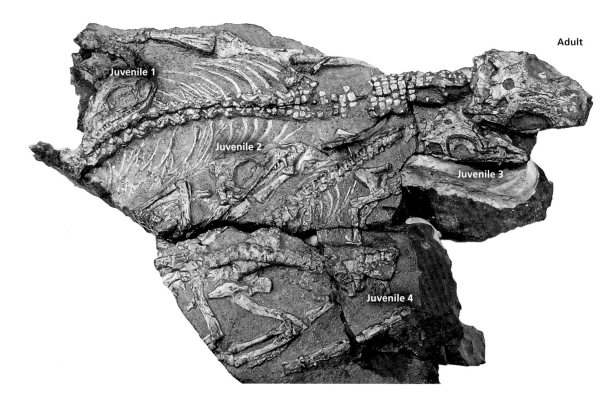

5.3. Permian social network! An adult varanopid (synapsid) fossilized with its offspring in a block, found in the Permian of South Africa (260 Ma) by my colleague Roger Malcom Smith. This assemblage is unique and is the oldest evidence of parental care in tetrapod evolution.

With the kind permission of Roger Malcom Smith (Iziko Museum of South Africa, Cape Town) and Jennifer Botha-Brink (National Museum of South Africa, Bloemfontein).

In addition to the geological background necessary to find fossils in the field and the biological savoir faire to identify and reconstruct them, a paleontologist must also have a grasp of physics. This is essential for gathering precious information about the locomotion of a fossil specimen using biomechanics—a discipline that relies on modeling, in three dimensions if possible—to demonstrate the way a bone articulates in its socket. First, the bones are scanned (using a surface or X-ray scanner, depending on available technology) and then reconstructed into 3-D digital images. These are then fed into a biomechanical software program that allows researchers to instantly visualize each bone's movement. This approach relies, once again, on state-of-the-art technology and allows an understanding of how an appendage (limb or fin) or a vertebral column moves, the better to reconstruct the locomotion of the vertebrate under study. We saw its application in chapter 1 when we discussed the terrestrial or aquatic ways of life of early Devonian tetrapods. Biomechanics can be, of course, complemented by the study of fossil trackways and footprints—a separate discipline called paleoichnology, which gives us an insight into how an animal moved (its stride, speed, and so on).

The paleontologist can also model the diet of the animal under study by analyzing its teeth, but finite element analysis (FEA) proves a useful ally. This method, well known to industrial designers, is in fact an analysis of forces exerted on an object, whether a screw or a stegocephalian jaw. FEA transforms an object into virtual polygons to better calculate its resistance to mechanical stress. Zones subject to pressures great and

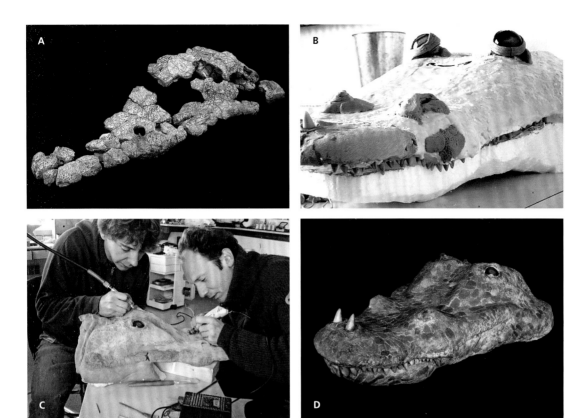

small are precisely identified and analyzed, allowing one, in the case of a jaw, to specify the animal's dietary behavior. This technique can also be applied to limb bones for a better understanding of locomotion (fig. 5.2).

Soft tissues, locomotion, diet — these are not the only reconstructions possible: in certain cases, the paleontologist even has access to the behavior of the animal they are studying. Fossil trackways and footprints sometimes reveal how an animal moved; taphonomic studies — the study of burial conditions and death — provide us, in cases of exceptional preservation, with fascinating information about the animal's behavior. My colleagues Jennifer Botha-Brink (Bloemfontein, South Africa) and Sean Modesto (Sydney, Canada) have recently shown that a small type of pelycosaur (varanopid) aged at 260 million years already displayed parental care. They found, in a block of argillite from the Middle Permian of South Africa, the skeleton of an adult — probably the mother — with those of four offspring around her (fig. 5.3). Taphonomic analysis suggests that this small family was surprised and rapidly buried by a mudflow. We already knew that, within the amniotes, some crocodiles and squamates (lizards and snakes), birds, mammals, and even some dinosaurs look (or looked) after their young. With this one-in-a-million discovery, the exceptional preservation of which has allowed the "fossilization of behavior," we can now be almost certain that parental care appeared very early in amniote history.

5.4. From fossil to 3-D reconstruction. *Nigerpeton ricqlesi* (see chapter 3), an amphibious temnospondyl from the Permian of Niger (discovered and described by the author), was a perfect candidate for anatomical reconstitution. Using different available fossils, such as a well preserved half-skull (A), a scale model in polyester resin is sculpted (B), by the model maker and sculptor Franck Limon-Duparcmeur (alias "the Dude," on the left, photo [C], aided by the author). Each scale, for example, was sculpted using a micro-drill, respecting the ornamentation visible on the original bones. The result (D) is remarkable!

Photos: reproduced with the kind permission of Christian Sidor (University of Washington; A) and Franck Limon-Duparcmeur (B–D).

A Brief Guide to Paleontology

The paleontologist tries to go through everything with a fine-toothed comb; nothing is left to chance. Even skin – notably its texture – can sometimes be reconstructed, when it has been miraculously preserved (epidermal remains of dinosaurs have been found, for example) or, as is more often the case, based on skeletal ornamentation (fig. 5.4) and comparisons with living animals. This illustrates the concept of actualism (also called uniformitarianism): in the living world, a tetrapod's skin is often of neutral tones, matching its environment. But when it lives in herds, its exterior can be heterogeneous (like that of a zebra or giraffe) and even highly colored if it has an intense social life (as do birds, for example). In aquatic environments, the ventral side of vertebrates is usually darker than the dorsal side, making it less visible to potential predators above and below. The careful paleontologist is not prey to overimagination.

A Bone, a Life

The external morphology of a bone is already a goldmine of information for paleontologists. But there is more to be gleaned: the study of inner bone structure, known as bone paleohistology, can also yield important clues about an extinct animal's lifestyle and habitat. Each bone, in fact, contains a structure that grows during its development: this structure also records constraints from the external environment (such as climate) as well as pressures acting directly on the skeleton during locomotion. A terrestrial tetrapod has a femoral diaphysis (central region of the femur) different from that of a flying or aquatic tetrapod (fig. 5.5).

How can we observe the inner structure of a bone? Today the technique consists of direct scanning using X-ray or proton microtomography (if using protons, the rays must be produced from a particle accelerator or synchrotron). The old-school approach consists of sectioning the fossil with a diamond saw to obtain thin slices. These are then polished using an abrasive that allows light to penetrate and reveal the inner structure of the bone, which is visible under the microscope (using natural or polarized light; fig. 5.6). The scanner is the expensive option, of course, but

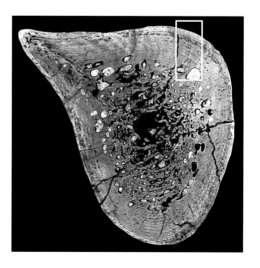

5.5. Diaphyseal section of a femur belonging to an adult temnospondyl from the Triassic of Morocco (215 Ma) and showing a typical aquatic structure: the section is very spongious at the center, becoming gradually more compact toward the periphery. The expansion of the bone on the top left corresponds to a typical femoral apophysis: the adductor ridge (see fig. 5.7 for a close-up of the boxed section).

Photo: Sébastien Steyer (CNRS/Muséum National d'Histoire Naturelle).

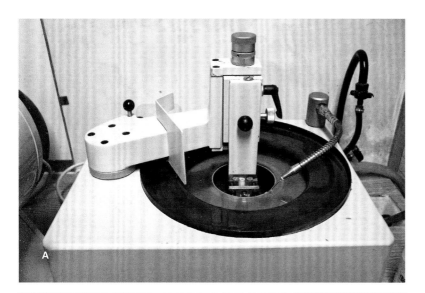

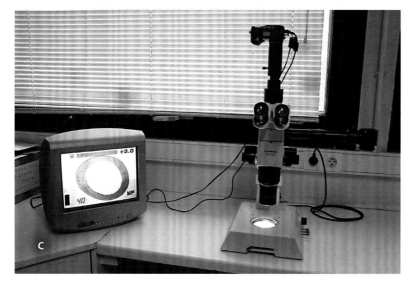

5.6. Bone paleohistology—getting at the inner bone structure. The old-school approach consists of sectioning the bone using a diamond saw (A) before being polished (B, the bone is in vertical position) until it is thin enough to allow light to penetrate. (C) The section is then analyzed using an optical microscope. Today, this technique, very useful for determining life-history traits of both extinct and extant organisms, is being replaced by X-ray (or proton) microtomography, which allows the sample to be scanned without destroying it.

Photo reproduced with the kind permission of Sophie Sanchez (Muséum National d'Histoire Naturelle).

A Brief Guide to Paleontology

5.7. Zoom into the cortex (the most external region) of a femoral section belonging to the same temnospondyl as in fig. 5.5. Several dark concentric lines are visible; these are lines of arrested growth (LAGs). They get closer together toward the periphery, indicating initial rapid growth followed by a slowdown with age.

Photo: Sébastien Steyer (CNRS/Muséum National d'Histoire Naturelle).

it does not destroy the fossil and one can obtain sections in every plane (transversal, longitudinal, and so on).

Some sections reveal dark concentric rings, or annuli, which are called lines of arrested growth (LAGs); they are a little like the growth rings in a tree trunk. These LAGs indicate temporary slowdowns in growth when the animal was going through difficult periods. In most living tetrapods, a LAG is formed every year, during estivation or hibernation. Consequently, the greater the distance between LAGs in a bone section, the faster the animal's growth (fig. 5.7), and the closer together the LAGs, the slower the growth. This is the principle of skeletochronology, a discipline that allows us to measure the rate and rhythm of growth in a given species, extant or extinct.

5.8. A mass spectrometer. This instrument analyzes matter in terms of the atomic mass of its components. Indispensable for isotopic biogeochemistry studies, it allows paleontologists to measure isotopic ratios of some elements (carbon, nitrogen, oxygen) in a fossil. This can yield valuable information (the animal's position in the food chain, paleoenvironmental conditions, etc.).

Photo reproduced with the kind permission of Fred Quillévéré (Université Claude-Bernard, Lyon, France).

5.9. Proportion of the isotope oxygen-18 (18O) in several living reptiles measured by the $\partial 18O$ (in ‰, after Lecuyer et al. [1999]). The 18O of the marine turtle *Dermochelys coriacea* and the marine crocodile *Crocodylus rhombifer* is around 3‰ greater than that of their freshwater counterparts. The Saharan viper's $\partial 18O$ is one of the highest. It would be interesting to superpose values corresponding to ichthyan sarcopterygians and early tetrapods onto this scale and perhaps help paleontologists to identify what type of environment—freshwater or marine—they lived in.

This approach was applied, for the first time, to the transition between ichthyan and tetrapod sarcopterygians (see chapter 1) by Sophie Sanchez in her doctoral research at the Muséum National d'Histoire Naturelle, which I co-supervised. The objective was to better delimit where these evolutionary events played out, as well as to identify life-history traits of the early tetrapods. Sanchez also analyzed several specimens of the seymouriamorph *Discosauriscus* (see chapter 4), revealing a growth rhythm more like that of a salamander than of a modern lizard.

Bone paleohistology and skeletochronology were developed in France in the 1970s and 1980s, thanks notably to Armand de Ricqlès of the Collège de France, and Jacques Castanet of l'Université Paris 6. As modern microtomography becomes more widely used in paleontology, disciplines such as this are on the rise.

A Whiff of Oxygen

Before we sing the praises of fieldwork, we should mention a last analytical method that allows us to resolve intriguing problems: isotopic biogeochemistry is the analysis of one or several isotopes and their relative

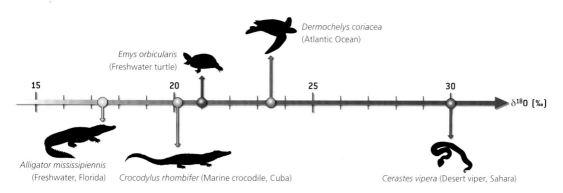

5.10. Extraction of a vertebral column of a pareiasaur (parareptile) discovered in the Permian deposits of the Nigerien Sahara (see chapter 3). After delicate cleaning with a brush (A), a systematic excavation is carried out all around the fossil to detect any further remains (B). The fossil and its surrounding matrix are then covered in a protective coat of plaster (C). Once dry, the cocoon is freed from the bedrock with a pick wielded by Christian Sidor (D). The plaster block is then turned over and transported by four-wheel drive (E) to the nearest preparation laboratory—in this case, several thousand kilometers away!

Photos: with the kind permission of Christian Sidor (University of Washington; A); Sébastien Steyer (CNRS/ Muséum National d'Histoire Naturelle; B–E).

proportions in a fossil. This discipline hinges on the fact that isotopes possess the same number of protons and electrons, but the number of neutrons varies.

When well preserved, fossils—notably teeth and skeletons—keep the original isotopic composition of the chemical elements extracted from their habitat to build their bodies. For example, the ratios of carbon isotopes ^{12}C and ^{13}C and of azotes ^{15}N and ^{14}N can allow us to position a fossil in its food chain and its paleoenvironmental conditions (For example, did it live in an enclosed habitat like a forest, or an open one like a savannah?). These isotopic ratios are calculated using mass spectrometry, a technique used to analyze matter in terms of the atomic mass of its components (fig. 5.8), and which is employed in paleontology using tiny samples of fossil tissue (such as the hydroxyapatite of a tooth). Before undertaking such an analysis, the paleontologist looks for any structural and chemical modifications that the fossil has undergone (diagenesis), to ensure that its state of preservation is suitable for an isotopic biogeochemistry approach—which is rarely the case!

Isotopic analysis of the oxygen is also rich in information. The proportion of isotope oxygen-18 in an organism's skeleton (measured by the $\partial^{18}O$) depends on its lifestyle and environment. For example, in the living world, a marine crocodile (or marine turtle) has a ^{18}O about 3‰ greater than that of a freshwater crocodile (or turtle)—an infinitesimally small difference that is, nevertheless, quantifiable (fig. 5.9). The desert viper's $\partial^{18}O$ is amongst the highest known (30‰, as opposed to 23.4‰ for the marine turtle *Dermochelys coriacea*, and 20.1‰ for the marine crocodile *Crocodylus rhombifer*). These differences seem to be linked to the strategies used to control salt levels in the body and reduce osmotic stress: the desert viper suffers a high rate of water loss and has to maintain high salt levels to survive.

The application of all these biochemical approaches on fossil remains (after having successfully passed the diagenetics test, of course) is very promising: it will perhaps allow us to better pinpoint the habitats of fossil organisms that haunted the margins between fresh and saltwater—as did many stegocephalians (see chapter 3). Ongoing studies will surely reveal fascinating facts about the paleoenvironments of these early tetrapods.

In Praise of Fieldwork

We have just sketched out the numerous approaches paleontologists employ to identify more precisely the morphology, physiology, and ecology of a fossil organism, even in a fragmentary state. But these analyses all depend on available fossil material. For this there are two solutions: spend your life rummaging through fossil collections, or get out into the field. The first solution is easier, but most important for a paleontologist and the lifeblood of studies is research in the field.

Paleontological fieldwork comprises both prospecting and excavating. Prospecting is carried out in targeted zones where the geology is known, outcrops on the surface (such as stone-covered plains) being

A Brief Guide to Paleontology

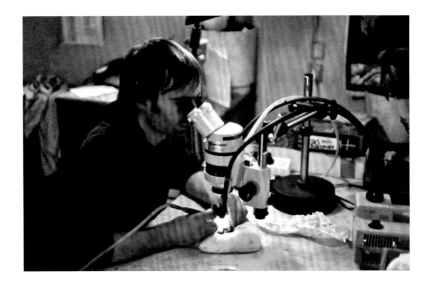

5.11. A small fossil is extracted from its matrix. Renaud Vacant, talented preparator (CNRS, Muséum National d'Histoire Naturelle) is on the job, armed with a stereo microscope, an airscribe (in his right hand), and endless patience. Requiring painstaking precision, the preparation of a fossil can sometimes take several months.

Photo: Sébastien Steyer (CNRS/Muséum National d'Histoire Naturelle).

preferable to outcrops exposed in section (such as cliffs). It consists of prospecting the largest possible zone of exposure, scanning for the slightest clue (perhaps a shard of bone or a tooth). To stack the odds in one's favor (when the geology is well known), efforts are first concentrated on specific sedimentary beds (such as ancient river courses or oxbows, where dead tree trunks and cadavers are more likely to have been transported and deposited). The greater the concentration of bone shards on the surface, the greater the chance of finding a bone or complete skeleton below. A larger prospecting team also increases the chances of a find. Once the fossiliferous zone is located, excavation can commence. When a fossil is found, it is not touched, at least in the case of vertebrate remains. The precious object is delicately cleaned on the surface with a brush and a trench is dug all around it. The fossil and its matrix are then covered in a layer of wet newspaper (toilet paper also works admirably) and, most important of all, plaster. The paper prevents the plaster from sticking to the bone and the plaster protects the fossil and keeps it firmly in its matrix. Once the plaster is dry, the cocoon is transported to the nearest preparation laboratory. This can entail a journey of several thousand kilometers (if you happen to be in the Niger desert, for example; fig. 5.10). It is only then that the matrix can be carefully removed using a pneumatic dentist's drill or a sandjet airscribe, sometimes millimeter by millimeter under a microscope or stereo microscope so as not to damage the fossil (fig. 5.11). Requiring the precision of a watchmaker, the work can take several months if the matrix is hard or the fossil complex. When the preparation is completed, the fossil is ready to be studied.

We must not forget that paleontology is a natural science. It is the Earth that gives up the raw materials on which all further analyses depend. Without fieldwork there would be no fossils and, without fossils, no paleontologists! It is therefore vital to promote fieldwork in addition to laboratory research; yet travel funds are getting thinner and thinner on the ground. The way things are going, the raw materials of research could become a rare commodity; this is exacerbated by an overreliance on modeling. Biochemist James E. Lovelock raises the alarm in *Gaia: A New Look at Life on Earth*:

> Unfortunately, most scientists live their lives in cities and have little or no contact with the natural world. . . . [T]hose of us who go forth in ships or travel to remote places . . . are few in number compared with those who choose to work in city-based institutions and universities. Personal contact between the explorers and the model builders is rare and information passes through the tense limited phraseology of scientific papers, where subtle, qualifying observations cannot be included among the data. (1979: 136–137)

Today, the search for fossils starts with a long and convoluted search for funding. The paleontologist must constantly remind funding bodies that to understand nature, past and present, you need to frequent it first.

BIBLIOGRAPHY

1. The Great Transition

BOOKS

Becker, R. T., and W. T. Kirchgasser (eds.). 2007. *Devonian Events and Correlations*. Geological Society Special Publication 278. London: Geological Society.

Bemis, W. E., W. W. Burggren, and N. E. Kemp (eds.). 1987. *The Biology and Evolution of Lungfishes*. New York: Alan R. Liss.

Clack, J. A. 2002. *Gaining Ground: The Origin and Evolution of Tetrapods*. Bloomington: Indiana University Press.

Darwin, C. 1859. *On the Origin of Species by Means of Natural Selection, or the Preservation of Favored Races in the Struggle for Life*. London: John Murray.

Foreman, R. E. D., A. Gorbman, J. M. Dodd, and R. Olsson. 1985. *Evolutionary Biology of Primitive Fishes*. New York: Plenum.

Forey, P. L. 1998. *History of the Coelacanth Fishes*. London: Chapman and Hall.

Friend, P. F., and B. Williams (eds.). 2000. *New Perspectives on the Old Red Sandstone*. Geological Society Special Publication 180. London: Geological Society.

Gould, S. J. 1991. *Bully for Brontosaurus*. New York: Norton.

Janvier, P. 2003. *Early Vertebrates*. Oxford: Clarendon Press.

Jarvik, E. 1980. *Basic Structure and Evolution of Vertebrates*. 2 vols. London: London Academic Press.

Long, J. A. 1995. *The Rise of Fishes: 500 Million Years of Evolution*. Baltimore: Johns Hopkins University Press.

Maisey, J. G. 1996. *Discovering Fossil Fishes*. New York: Henry Holt.

Ricqlès, A. de. 1998. Les fossiles vivants n'existent pas. In H. Le Guyader (ed.), *L'evolution*, pp. 78–83. Paris: Belin–Pour la Science.

Schultze, H.-P., and L. Trueb (eds.). 1991. *Origins of the Higher Groups of Tetrapods: Controversy and Consensus*. New York: Comstock.

Smith, J. L. B. 1956. *Old Fourlegs: The Story of the Coelacanth*. London: Longmans.

Steyer, S. 2001. Les amphibiens, premiers vertébrés terrestres. In Sélection du Reader's Digest (ed.), *La fabuleuse histoire de la Terre*, pp. 99–103. Paris: Reader's Digest.

ARTICLES

Ahlberg, P. E. 1989. Fossil fishes from Gogo. *Nature* 337: 511–512.

Ahlbert, P. E., and J. A. Clack. 2006. A firm step from water to land. *Nature* 440: 747–749.

Arsenault, M., H. Lelièvre, and P. Janvier (eds.). 1995. Proceedings of the 7th International Symposium on Lower Vertebrates. *Bulletin du Muséum national d'histoire naturelle de Paris*, 4th ser., 17C.

Boisvert, C. 2005. The pelvic fin and girdle of *Panderichthys* and the origin of tetrapod locomotion. *Nature* 438: 1145–1147.

Clack, J. A. 2006. Le premier pied à terre. *Pour la Science* 340: 30–36.

Clément, G. 2006. Les tétrapodes de paléo-Belgique. *Pour la Science* 340: 38–43.

Niedzwiedzki, G., P. Szrek, K. Narkiewicz, M. Narkiewicz, and P. E. Ahlberg. 2010. Tetrapod trackways from the early Middle Devonian period of Poland. *Nature* 463: 43–48.

Schultze, H.-P., and M. Arsenault. 1985. The panderichthyid fish *Elpistostege*: A close relative of tetrapods? *Palaeontology* 28: 293–309.

Steyer, S. 2007. Premiers pas sur la Terre, grand pas pour l'évolution. *Bulletin d'Information de la Société des Amis du Muséum* 229: 5.

2. Limbs

BOOKS

Bonner, J. T. (ed.). 1982. *Evolution and Development*. Berlin: Springer-Verlag.

Gould, S. J. 1977. *Ontogeny and Phylogeny*. Cambridge, Mass.: Harvard University Press.

———. 1991. *Bully for Brontosaurus*. New York: Norton.

———. 2002. *The Structure of Evolutionary Theory*. Cambridge, Mass.: Harvard University Press.

Haeckel, E. 1866. *Generelle Morphologie der Organismen.* Berlin: G. Reimer.

———. 1868. *Natürliche Schöpfungsgeschichte: Gemeinverständliche wissenschaftliche Vorträge über die Entwickelungslehre im Allgemeinen und diejenige von Darwin, Goethe, und Lamarck im Besonderen, über die Anwendung derselben auf den Ursprung des Menschen und andere damit zusammenhängende Grundfragen der Naturwissenschaft* [The history of creation, or the development of the Earth and its inhabitants by the action of natural causes: A popular exposition of the doctrine of evolution in general, and that of Darwin, Goethe, and Lamarck in particular]. Berlin: G. Reimer.

Hinchliffe, J. R., J. M. Hurle, and D. Summerbell. 1991. *Developmental Patterning of the Vertebrate Limb.* New York: Plenum Press.

Mundson, R. 2007. *The Changing Role of the Embryo in Evolutionary Thought: Roots of Evo-Devo.* Cambridge: Cambridge University Press.

Muséum National d'Histoire Naturelle (ed.). 1991. *On a marché sur la Terre: Le roman de l'évolution.* Paris: ICS–Muséum National d'Histoire Naturelle.

Owen, R. 1849. *On the Nature of Limbs: A Discourse Delivered on Friday, February 9, at an Evening Meeting of the Royal Institution of Great Britain.* London: John Van Voorst.

Schopf, T. J. M. (ed.). 1985. *Models in Paleobiology.* San Francisco: Freeman, Cooper.

Shubin, N. 2008. *Your Inner Fish: A Journey into the 3.5 Billion-Year History of the Human Body.* New York: Random House.

ARTICLES

Duboule, D., and P. Sordino. 1997. L'origine des doigts. *La Recherche* 296: 66–69.

Steyer, S. 2000. Révolutions dans l'évolution. *Revue du Palais de la découverte* 274: 54–55.

———. 2000. Les pattes des amphibiens, entre bricolage et innovation. *La valse des espèces*, special issue, *Pour la Science* 28: 54–59.

Wu, X.-C., Z. Li, B.-C. Zhou, and Z.-M. Dong. 2003. A polydactylous amniote from the Triassic period. *Nature* 426: 516.

Zacany, J., and D. Duboule. 2007. The role of Hox genes during vertebrate limb development. *Current Opinion in Genetics and Development* 17: 359–366.

3. The Pangaean Chronicles

BOOKS

Anderson, J., and H.-D. Sues (eds.). 2007. *Major Transitions in Vertebrate Evolution.* Bloomington: Indiana University Press.

Arratia, G., M. V. H. Wilson, and R. Cloutier (eds.). 2004. *Recent Advances in the Origin and Early Radiation of Vertebrates.* Munich: Friedrich Pfeil.

Carroll, R. L. 1988. *Vertebrate Paleontology and Evolution.* New York: W. H. Freeman.

Czerkas, S. J., and S. A. Czerkas. 1990. *Dinosaurs: A Complete World History.* London: Dragon's World.

Gould, S. J. 1987. *Urchin in the Storm.* New York: Norton.

Gould, S. J. (ed.). 1993. *The Book of Life: An Illustrated History of Life on Earth.* New York: Norton.

Heatwole, H., and R. L. Carroll (eds.). 2000. *Palaeontology: the Evolutionary History of Amphibians.* Vol. 4 of *Amphibian Biology.* London: Beatty and Sons.

Hubert, A. B., and W. N. Schwarze. 1910. *David Zeisberger's History of the Northern American Indians.* Columbus: Ohio State Archaeological and Historical Society.

Laurin, M. 2008. *Systématique, paléontologie et biologie évolutive moderne: L'exemple de la sortie des eaux chez les vertébrés.* Paris: Ellipses.

Laurin, M., and R. R. Reisz. 1997. A new perspective on tetrapod phylogeny. In S. S. Sumida and K. L. M. Martin (eds.), *Amniote Origins: Completing the Transition to Land*, pp. 9–59. New York: Academic.

Lichtenberg, G. C. 1997. *Le miroir de l'âme: Aphorismes.* Trans. Charles Le Blanc. Paris: J. Corti.

Nitecki, H. (ed.) 1979. *Mazon Creek Fossils.* New York: Academic.

Poplin, C., and D. Heyler (eds.). 1994. *Quand le Massif Central était sous l'Équateur: Un écosystème carbonifère à Montceau-les-Mines.* Paris: CTHS.

ARTICLES

Carroll, R. L. 2007. The Palaeozoic ancestry of salamanders, frogs and caecilians. *Zoological Journal of the Linnean Society* 150: 1–140.

Lombard, R. E., and S. S. Sumida. 1992. Recent progress in the study of early tetrapods. *American Zoologist* 32: 609–622.

Milner, A. R., and S. E. K. Sequeira. 1994. The temnospondyl amphibians from the Viséan of East Kirkton, West Lothian, Scotland. *Transactions of the Royal Society of Edinburgh* 84: 331–361.

Ruta, M., M. I. Coates, and D. L. J. Quicke. 2003. Early tetrapod relationships revisited. *Biological Reviews* 78: 251–345.

Sidor, C. A., F. R. O'Keefe, R. Damiani, S. Steyer, R. M. H. Smith, H. C. E. Larsson, P. C. Sereno, O. Ide, and A. Maga. 2005. Permian tetrapods from the Sahara show climate-controlled endemism in Pangaea. *Nature* 434: 886–889.

Strother, P. K., L. Battison, M. D. Brasier, and C. H. Wellman. 2011. Earth's earliest non-marine eukaryotes. *Nature* 473: 505–509.

4. Between Earth and Sky

BOOKS

Bakker, R. T. 1986. *The Dinosaur Heresies : New Theories Unlocking the Mystery of the Dinosaurs and Their Extinction.* New York: William Morrow.

Benton, M. J. (ed.). 1988. *Amphibians, Reptiles, Birds.* Vol. 1 of *The Phylogeny and Classification of the Tetrapods.* Oxford: Clarendon Press.

Benton, M. J. 2003. *When Life Nearly Died Out: The Greatest Mass Extinction of All Time.* London: Thames and Hudson.

Buffetaut, E. 2003. *La fin des dinosaures: Comment les grandes extinction ont façonné le monde vivant.* Paris: Fayard.

Erwin, D. H. 1993. *The Great Paleozoic Crisis: Life and Death in the Permian.* New York: Columbia University Press.

Raup, D. M. 1997. *De l'extinction des espèces: Sur les causes de la disparition des dinosaures et de quelques milliards d'autres.* Paris: Gallimard.

Sepkovski, J. J. 1996. Patterns of Phanerozoic extinction: A perspective from global data bases. In O. H. Walliser (ed.), *Global Events and Event Stratigraphy in the Phanerozoic,* pp. 35–51. Berlin: Springer-Verlag.

Stümpke, H. 1962. *Anatomie et biologie des Rhinogrades, un nouvel ordre de Mammifères.* Paris: Masson.

Sumida, S. S., and K. L. M. Martin (eds.). 1997. *Amniote Origins: Completing the Transition to Land.* New York: Academic.

ARTICLES

Gand, G., J. Garric, G. Demathieu, and P. Ellenberger. 2000. La palichnofaune de vertébrés tétrapodes du Permien Supérieur du bassin de Lodève (Languedoc, France). *Palaeovertebrata* 29: 1–82.

Rybczynski, R., and R. R. Reisz. 2001. Earliest evidence for efficient oral processing in a terrestrial herbivore. *Nature* 411: 684–687.

Steyer, S. 2006. Quand la vie a failli disparaître: Un scenario de crise il y a 250 millions d'années. *Cosinus* 73–74: 18–23, 29–35.

———. 2007. Le Permien au Sahara. *La Recherche* 409: 36–41.

5. A Brief Guide to Paleontology

BOOKS

Cuvier, G. 1812/1992. *Recherches sur les ossements fossiles de quadrupèdes: Où l'on établit les caractères de plusieurs espèces d'animaux que les révolutions du globe paraissent avoir détruites: Discours préliminaires.* Ed. Pierre Pelligrin. Rev. ed., Paris: Flammarion.

Lécuyer, C., P. Grandjean, J.-M. Mazin, and V. de Buffrénil. 1999. Oxygen isotope compositions of reptile bones and teeth: A potential record of terrestrial and marine paleo-environments. In E. Hoch and A. K. Brantsen (eds.), *Secondary Adaptation to Life in Water,* p. 78. Copenhagen: Geologisk Museum.

Lovelock, J. E. 1979. *Gaia: A New Look at Life on Earth.* Oxford: Oxford University Press.

ARTICLE

Steyer, S. 2005. Voyages naturalistes en crise. *Cadmos* 8: 59–69.

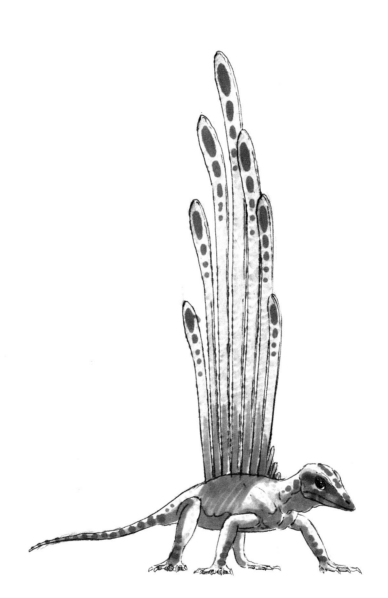

INDEX

Acanthostega, 1–2, 11, 13, 15, 18–19, 22–23, 25–26, 28, 32–36, 43, 47, 51, 53–55, 59, 61, 64, 114
adaptation, 25, 56–57, 65, 74, 78, 81, 133, 137
AER (apical ectodermal ridge) formation, 45–47, 52
aistopod, 93, 98–99, 106
Alland, William, 8
alligator, 22–23, 108, 129, 169
Alvarez, Luis, 146
amber, 40
amniote, xii, 53, 55, 59–61, 64–65, 92, 107, 111–128, 132–140, 143–144, 147–148, 154–157, 165
Amphibamus, 76–78
amphibian, xi, 1, 8, 24, 30, 35, 38, 55, 59–63, 65, 68–101, 105–106, 108–109, 111, 127, 144, 157. *See also* tetrapod
Anderson, Jason, 60, 105
Andrias, 103, 108–109
anoxia, 63, 145
Antarctica, 60, 74, 86, 146, 155
anthracosaur, 111–116
anthropocentric, 1, 56
Aphelosaurus, 123, 130–135
apoptosis, 47–48, 52, 101
araeoscelid, 131–133
Archaeopteris, 38–39
Archaeopteryx, 39, 135
Archaeothyris, 123–124, 126–127
Archeria, 112
Arctic, 17, 60
Argana Basin (Morocco), 95, 158
Argentina, 157
Arsenault, Marius, 15
Arthropleura, 67
arthropod, 5, 20, 63, 65–67, 98, 108, 125, 140, 144
Australia, 6–7, 26, 28–29, 60, 86, 141, 146
axis (anatomy), 46–47, 50–52
axolotl, 97, 100

batrachomorph, 61, 65, 92, 111
Battail, Bernard, 81

Beagle, 101, 125
Bedout Crater, 146
Belgium, 26–28
benthic (organism), 15, 22, 63, 93, 145
Benton, Michael, 111, 143
Berman, Dave, 120, 134
biodiversity, xi, 40, 56, 143–144
biomechanical, 35, 40, 55, 164
Birgus, 65
blastoid, 144
Boisvert, Catherine, 15
Bolt, John, 76, 78
Boskovice Basin (Czech Republic), 118, 120
Botha-Brink, Jennifer, 164–165
Brachydectes, 99, 100
Brazilosaurus, 128
Buffetaut, Eric, 143
Bunostegos, 90, 92

caecilian, 59–62, 98, 101, 105–107
camouflage, 55, 70, 93, 153
Canada, 71–73, 124–126, 165
Captorhinus, 122
Carroll, Robert, 60
cast, 11–12, 78, 104, 115
Castanet, Jacques, 169
chance, 2, 81, 135, 166
character, 2, 4–7, 9, 11, 13–15, 18, 20–23, 25, 27, 30, 35, 43, 59–60, 62, 74, 76–77, 92–93, 97–99, 101–106, 108–109, 111–112, 114, 116–118, 120–123, 125–126, 128, 133, 135–136, 138, 141, 147–150, 156–157
Chicxulub Crater, 146
chimera, 24
China, 26–28, 53, 55, 64, 66, 74, 103–104, 108, 142–143, 145, 155
chiridian (limb), 1–3, 5, 7, 9, 12, 15, 20, 22, 24–25, 31–32, 34, 43–44, 48–57, 59, 64, 74
Chroniosaurus, 112
Chrysopelea, 138
Chunerpeton, 103–105, 108–109
Clack, Jennifer, xiii, 24, 31–35, 54

clade, xii, 6, 9, 22, 59–61, 65, 89, 101, 106, 111–112, 123–125, 141, 148–149
cladistics, 62. *See also* phylogenetic analysis
cladogram, 21, 59–61, 71, 89
claw, 130–131, 134–135, 149–153, 161, 163
Clément, Gaël, xiii, 27–28
climate, 63, 65, 92, 145, 147, 166
coelacanth, xii, 3–11, 44, 48, 109
Coelurosauravus, 111, 123, 135–138
comparative anatomy, 5, 108, 161
continental drift, 128–129
Courtney-Latimer, Marjorie, 3
Crassigyrinus, 61, 68–72
crisis (biological), 63–64, 83, 107, 143–147, 156–158
crocodile, xii, 18, 55, 70, 74, 76, 81, 86, 111–112, 114, 117, 123, 144, 157, 165, 169–170
Crocodylus, 169–170
Cryptobranchus, 108
Cuvier, Georges, 161
Cyclopes, 151–152
Cynocephalus, 137–138
Cynognathus, 118
Czech Republic, 65, 97, 117–118

Darwin, Charles, xii, 1, 101, 125
Dawson, William, 124–125
Deccan Traps (India), 146
Densignathus, 26–27
De Ricqlès, Armand, 82–83, 169
Dermochelys, 169–170
Devonian (period), 28, 63, 65, 74, 76, 78, 81–86, 88–95, 97–100, 103, 107, 111–121, 127–136, 138–149, 155–157, 161, 164–165, 170
diadectomorph, 60–61, 111–112, 116, 118, 120–121, 141
diagenesis, 170
Dicynodon, 142–143
dicynodont, 83, 139, 142, 156, 158
digit, xi–xii, 1–2, 6, 12–13, 18–19, 21–22, 24–25, 30–31,

179

34–36, 43–56, 59, 78, 111, 114, 128, 137, 149–150, 153
"dignathic heterodonty," 68
Dimetrodon, 74, 123, 138–141
dinosaur, xi–xii, 4, 10, 39, 43, 56, 60, 74, 82, 107, 111, 121, 123, 135, 138, 141, 143–144, 147–148, 151–152, 155–158, 162–163, 165–166
Diplocaulus, 93–95
dipnoi, 5–9, 30, 48, 99
Discosauriscus, 117–120, 169
dissorophoid, 76–77, 99
diversity, 25, 28, 61, 101, 114, 143–144, 155, 158
Draco, 137–138
Drepanosaurus, 123, 150–154
drosophila, 49
Duboule, Denis, 49–50, 52
double respiration, 6
Dutuit, Jean-Michel, 95

ecological niche, 65, 86, 107, 114, 136, 140, 144, 147, 154–155, 158
ecomorphotype, 109
ectotherm, 101, 138
Edinburgh, 69, 126
Edingerella, 78–81
eel, 6, 93
Elginerpeton, 26–28
Ellenberger, Paul, 81
Elpistostege, 15–18, 22
embryo, xi, 21, 24, 43, 50, 52, 111, 121
embryogenesis, 21, 52
Emeishan Traps (China), 145
endocranium, 10–11, 161
endotherm, 138
Eocaecilia, 105–107
Eryops, 74
Eudibamus, 123, 134–135
euryhalinity, 38, 80, 91
Eurypterus, 67
Eusthenopteron, 10–13, 15, 17–20, 22–25, 30, 43
evo-devo, 43–44, 48, 59
evolution, xi–xii, 1–2, 5–6, 8, 13, 21, 28–31, 43–45, 48, 53–57, 59–60, 62, 64–67, 69, 74, 76, 93, 95, 98–99, 103, 105–107, 111–112, 120–121, 123–125, 127, 133, 147–149, 151–152, 156–158, 164, 169
evolutionary contingency, 55
evolutionary convergence, 76, 93, 95, 106, 148, 158
evolutionary scenario, 31
exaptation, 56–57, 137–138

extinction, xi, 60, 63–64, 74, 107, 111, 114, 143–148

Falconnet, Jocelyn, xiii, 130, 133, 135
Famennian (age), 25–26, 30–32, 63–64, 143
fin, 1–15, 17–20, 22, 24–25, 30–31, 34–35, 43–44, 48–52, 54–56, 69, 164
finite element analysis (FEA), 163–164
Finney, Sarah, 34
Florence, 126
food chain, 92, 114, 157, 169–170
footprints, 29, 132, 135, 164–165
fossil (stratigraphic), 142
fossil (trace), 29, 67, 95, 99, 132, 135
fossilization, 40, 148–149, 155, 162
France, 40, 65, 80, 99, 117, 131–134, 163, 169
Frasnian (age), 17, 26, 29, 63–64, 143
Frasnian-Famennian crisis, 63–64, 143
function, 1, 25, 39, 50, 54–57, 93–94, 138, 149, 151, 153

Gand, Georges, xiii, 132, 135
gap, 10, 20, 28–30, 68, 127
gene (homeotic), 49–50
genetics, 21, 39, 43, 49–51, 61
Germain, Damien, xiii, 94–95, 98
Germany, 65, 82, 95, 116, 132–136, 155–158
Gerrothorax, 93, 95
Gervais, Paul, 133
gills, 6–9, 18, 23, 25, 30, 34, 68, 97, 100, 103–105, 108
Ginsburg, Léonard, 81
glaciation, 63
gland, 72, 80, 100–101; salt, 80
Glaucomys, 137–138
Gondwana, 28, 86, 89 (map), 90, 95
Gorgonopsian, 83, 90–92, 142
Gould, Stephen Jay, 3, 21, 55, 62, 95, 137
Grassé, Pierre-Paul, 147
greenhouse gas, 146
Greenland, 26, 28–34

Homo sapiens, 5, 137
Horton Bluff (Canada), 68
hox (gene), 49–52, 55
hybodont (shark), 80

Hylonomus, 123–127
Hynerpeton, 26–28, 36–38
Hypuronector, 123, 151–155

Iberospondylus, 78
ichthyan (fish-like forms), 1, 4–6, 9–16, 18–26, 30, 34–35, 38, 43, 49, 59–60, 64, 169
Ichthyostega, 2, 10–11, 13, 15, 18, 22–26, 28, 30–36, 42–43, 51, 53–55, 61, 64, 100, 114
Indian Ocean, 3, 146
innovation, 3, 21–22, 54–56, 59, 61, 66, 93, 121, 123–124
intracranial articulation, 11, 13
isotope, 169–170
Italy, 79–80, 149–152

Jakubsonia, 26, 28
Jalil, Nour-Eddine, xiii, 95
Janvier, Philippe, 28–30, 81
"Janvier's gap," 28–30
Japan, 108–109, 145
Jarvik, Erik, 10–12, 30, 32, 44
Joggins (Canada), 124–126

Karaurus, 103–105
Kazakhstan, 104–105, 145
Klembara, Jozef, 112, 117–118
Kyrgyzstan, 148, 155

labyrinthodont, 12, 60, 81, 93
Landry, Joseph, 10
Laos, 114, 142–143
Latimeria, 3, 5
Latvia, 26, 28
Laurasia, 28, 89 (map), 90, 95, 114, 116
Laurin, Michel, 60, 100
Lehman, Jean-Pierre, 82
Lepidosiren, 5–9
lepidotrich (dermal ray), 6, 12–14, 43–44, 48–49, 52, 54–55
lepospondyl, 60–61, 65, 76–77, 92–101, 105–106, 111, 127, 144
Lesotho, 81–82, 156–157
Limnoscelis, 118, 121
Limon-Duparcmeur, Franck, 86, 165
lissamphibian, 55, 59–62, 65, 74, 76–78, 93, 99–109, 111–112, 120
lizard, xii, 2, 97, 111–112, 117, 125–126, 134, 137–138, 149–150, 157, 165, 169
Lodève Basin (France), 131–133, 135

Lombard, R. Eric, 60
Longisquama, 123, 148–149, 151, 154–155
Lovelock, James E., 173
Lyell, Charles, 125
lysorophian, 99–100

Madagascar, 78–80, 101–102, 136, 167
Maganuco, Simone, 79–80
Majungasaurus, 162
mammal, xi–xii, 23–24, 61, 74, 107, 111, 121, 123, 125, 138, 147, 151, 155–157, 165
"mammal-like reptile," 24, 61, 74, 83, 91–92, 118, 120, 125–127, 140–141, 158
Mastodonsaurus, 82
mating, 42, 55, 108, 150
Mazon Creek (Illinois), 77
Megalancosaurus, 123, 149–152, 154
Meganeura, 67, 125
Megarachnides, 67
Menopoma, 108
mesosaur, 123, 127–129, 132, 144
Mesosaurus, 123, 127–129, 132
metamorphosis, 97, 100–101, 117
Metaxygnathus, 26, 28
meteorite, xi, 145–146
Microbrachis, 97
Microsauripus, 132, 135
microtomography, 11, 102, 161–162, 166–167, 169
micro-wear, 121
Miguasha (Québec), 10, 15, 17
Milner, Andrew, 68
mimicry, 153
Modesto, Sean, 165
monobasal articulation, 4, 6, 22, 25
Montceau-les-Mines (France), 99
Moradi (Niger), 59, 82–86, 91–92, 119, 161
Moradisaurus, 82–86, 91–92, 119, 161
Morocco, 94–95, 107, 158, 166
morphogenesis, xi, 44–45, 49–50, 52, 55, 99
morphology, xi–xii, 3, 5, 7, 9, 21, 30, 40, 61, 64, 68, 71, 80, 92–93, 95, 97, 100, 104–105, 108–109, 112, 114, 118, 134, 150–151, 156–157, 161, 166, 170
mosaic, 11, 13–14, 21, 59, 114, 116
mosasaur, 127, 129
mutant, 8, 43, 47 49, 51–52

natural selection, 140
Neoceratodus, 6
neoformation, 35, 48, 54
neoteny, 97, 109
New South Wales (Australia), 26
Niger, 59, 82–86, 88–92, 161, 165, 170, 172
Nigerpeton, 86, 89–91, 165
North America, 60, 64–65, 74, 76, 86, 93, 98, 106, 114, 116, 118–119, 121, 127, 133, 138, 151, 153, 155
Norway, 74, 145
"nuclear winter," 146
Nunavut Territory (Canada), 17
Nyrany (Czech Republic), 97

Obruchevichthys, 26, 28
Oligokyphus, 123, 155
ontogeny, 62
Ophiderpeton, 98
osmotic stress, 80, 170
otolith, 23
Otopteryx, 147
oxygen, 8–9, 63, 67, 169–170

Pachygenelus, 123, 156
paleoenvironment, 26, 39, 40, 54, 86, 169–170
paleohistology, bone, 161, 166–167, 169
paleoichnes, 135
paleoichnology, 164
panchronic species, 5–6, 109
Panderichthys, 13–16, 18, 22–23, 43, 51
Pangaea, 59, 63, 65, 74, 82–84, 86, 88, 89 (map), 90, 92, 95, 114, 121, 129, 132, 142, 145–146, 155
parallelism, 106
parasagittal limbs, 116, 120, 134
pelycosaur, 138–141, 144, 156–157, 165
Permian (period), 28, 63, 65, 74, 76, 78, 81–100, 103, 107, 111–121, 127–135, 138–149, 155–157, 161, 164–165, 170
Permian-Triassic crisis, 63, 83, 107, 114, 116, 143–147, 155–157
photosynthesis, 66, 146
phylogenetic analysis, 62, 71, 76, 89, 99, 114, 116, 148
phylogenetic tree, 65, 86, 101
phylogeny, xii, 13, 60–64, 74, 76, 78, 80, 99, 114, 116, 120, 148
placoderm, 5–6, 38
Plateosaurus, 158, 163

poikilotherm organism, 101, 138
polydactyly, 32, 35–36, 43, 47, 50, 52–55
polyphalangism, 53
Protopterus, 6–8

Québec, 10, 15

radiation (diversity), 25, 28–29, 60, 64–65, 74, 107, 126–127, 141, 147
Rage, Jean-Claude, xiii, 103
Red Hill (Pennsylvania), 26, 36, 38
Reisz, Robert, 60, 100, 124, 140
reptile, xi–xii, 1, 24, 35, 53–55, 61, 74, 82–84, 86, 90–92, 95, 97, 106–107, 117–118, 120, 123–127, 132–136, 138, 143, 154–155, 169–170
reptiliomorph, 61, 71, 111–112, 120
Rhacophorus, 137–138
rhinograde, 147–148
Romer, A. S. (Alfred Sherwood), 28, 123
Romer's Gap, 28, 68
Rubricacaecilia, 107
Russia, 13–14, 26, 28, 35–36, 40, 82–83, 104, 112–114, 140–143, 145, 148
Ruta, Marcello, 60, 76

Sahara, 74, 82, 84–86, 88–91, 161, 169–170
Saharastega, 86, 88–91
salamander, 60–61, 68, 74, 97, 100–101, 103–105, 108–109, 118, 120, 169
Sanchez, Sophie, 120, 167, 169
sarcopterygians (lobe-finned), 4–6, 9–13, 15–22, 24–25, 35, 43–44, 48–49, 51, 54, 57, 59–60, 109, 169
sauropsid, xii, 1, 123–127, 129–130, 132–136, 138, 140, 144, 147–148, 151, 153, 157
Säve-Söderbergh, Gunnar, 26, 30, 32
scanner, 11, 164, 166
Schultze, Hans-Peter, 15
Scleromochlus, 158
Scotland, 26–29, 66, 68–69, 72, 95, 124–126
Senter, Phil, 149–150
Seymouria, 60, 111–117, 120
seymouriamorph, 60, 111–112, 114–118, 120, 169

Sidor, Christian, xiii, 83, 85–86, 88, 91, 161, 165, 170
simiosaur, 149–151, 153–154
Sinostega, 26–28
Smith, J. L. B., 3–4
Smith, Roger Malcom, xiii, 85, 164
snake, xii, 1, 45, 70, 98–99, 106, 111, 123, 138, 157, 165
South Africa, 3, 82–85, 142–143, 146, 156–157, 164–165
South America, 6–7, 86, 127–129, 132, 157
Spathicephalus, 61, 68, 71–73, 93
Sphyrna, 93, 95
Spiloblatta, 67
squamate, 123, 149, 157, 165
stegocephalian, 24, 59–65, 68–69, 71–79, 86, 88, 92–93, 104, 109, 111–117, 120–127, 133, 141, 144, 164, 170
Steiner, Gerolf, 147
Stereosternum, 128
Suchomimus, 151–152
Sumida, Stuart, xiii, 60, 119
Suminia, 123, 140–141
synapsid, 61, 74, 83, 90–92, 120, 123–127, 138–144, 147, 155–157, 164

Talpa, 151–152
talpid (mutant), 43, 47
Tanzania, 142–143
taphonomy, 40
Taquet, Philippe, 82–83

temnospondyl, 60–61, 65, 68–70, 74, 76–95, 99, 101, 111, 120, 125, 127, 144, 165–166, 168
Tersomius, 122
tetrapod, xi–xii, 1–39, 43–45, 48–56, 59–64, 68–74, 80, 82–83, 93, 98, 100–101, 111, 116–121, 123, 125–128, 132–133, 135–144, 149, 151–153, 156, 164, 166–170; amniote origins, 121–123; aquatic (marine), 2–3, 13, 26, 53–55, 63–64, 78–81, 91, 127–129, 143–145, 169–170; bipedal, 132–133; fish-tetrapod transition, 48, 51; flying, 135–139; locomotion, 2, 24, 35, 68, 109, 112, 117–119, 121, 152, 161, 164–166; non-amniote origins, 4–25, 35, 51; terrestrial, 1–2, 15, 19, 22, 25, 30, 32, 36, 40, 63–70, 74, 76–78, 80, 82, 92–93, 98–100, 112, 117–121, 125–126, 133, 135–136, 139–143, 150, 153–154, 164, 166. *See also* amphibian
tetrapodomorph, 4–5, 9–10, 13–23, 25–26, 34, 59
thermoregulation, 57, 137–138
Thévenin, Armand, 133
Ticinosuchus, 158
Tiktaalik, 5, 13, 17–23, 51
trace fossil, 29, 67, 95, 99, 132, 135
traps (geology), 145–146
tree fern, 20
trees (arboreality), 38, 68, 70, 92, 125, 135, 140, 151, 168, 172

Triadobatrachus, 101–103, 107
Triassic (period), 53, 55, 63, 65, 74, 78–83, 93, 95, 101–102, 107, 111, 114, 116, 135, 141, 143–158, 163, 166
trilobite, 63, 144
trithelodont, 155–157
tritylodont, 155–157
trophic spectrum, 120
Tulerpeton, 13, 26, 28, 35–36, 40, 53
tympanum, 23, 78, 114, 120

United Kingdom, 68–69, 71–73, 111, 136, 145, 155, 158
United States, 26–28, 36, 74, 76–77, 94, 99, 107–108, 116, 118–119, 155

Vacant, Renaud, xiii, 95, 172
Ventastega, 26, 28
Vianey-Liaud, Monique, 132

Wegener, Alfred, 128–129
Westlothiana, 126
Westoll, Stanley, 15
Wood, Stan, 68, 126

X-ray, 102, 161, 164, 166–167

Zeisberger, David, 108
Zenaspis pagei, 93, 95
ZPA (zone of polarizing activity), 46–47